Differential Equations: An Introduction

Differential Equations: An Introduction

Alexis Fletcher

MURPHY & MOORE
www.murphy-moorepublishing.com

Published by Murphy & Moore Publishing,
1 Rockefeller Plaza,
New York City, NY 10020, USA

ISBN: 978-1-63987-162-9

Cataloging-in-Publication Data

Differential equations : an introduction / Alexis Fletcher.
p. cm.
Includes bibliographical references and index.
ISBN 978-1-63987-162-9
1. Differential equations. 2. Calculus. 3. Bessel functions. I. Fletcher, Alexis.
QA371 .D54 2022
515.35--dc23

For information on all Murphy & Moore Publications
visit our website at www.murphy-moorepublishing.com

MURPHY & MOORE

Table of Contents

Preface

The purpose of this book is to help students understand the fundamental concepts of this discipline. It is designed to motivate students to learn and prosper. I am grateful for the support of my colleagues. I would also like to acknowledge the encouragement of my family.

In mathematics, differential equations refer to the study of equation that relates to one or more functions and their derivatives. These functions represent physical quantities, the derivatives show the rate of change and the equation itself defines the relationship between the two. The field is concerned with the study of solution of these equations and their properties. They can be divided into many types. Some of these are partial differential equation, ordinary differential equations, non-linear differential equations, and equation order. They are also classified as linear and homogeneous differential equations. Differential equations play an important role in other disciplines such as applied and pure mathematics, physics, biology, economics, and engineering. The book aims to shed light on some of the unexplored aspects of differential equations. It presents the complex subject of differential equations in the most comprehensible and easy to understand language. This book will serve as a reference to a broad spectrum of readers.

A foreword for all the chapters is provided below:

Chapter - Understanding Differential Equation

Differential equation can be termed as a relationship between the derivatives of one or more functions. The degree of a differential equation is the highest power of the order derivative and the order of a differential equation refers to the highest number of derivatives in an equation. This is an introductory chapter which will briefly introduce about differential equations.

Chapter - Ordinary Differential Equations

The type of differential equation that contains one or more functions of an independent variable is termed as ordinary differential equation. First-order ordinary differential equation, second-order ordinary differential equation, Abel's identity, Bernoulli differential equation, etc. are studied under its domain. The topics elaborated in this chapter will help in gaining a better perspective about ordinary differential equations.

Chapter - Linear and Non-Linear Differential Equations

Any differential equation which is defined by a linear polynomial in the unknown function is known as a linear differential equation. A differential equation that is not a linear equation in the unknown function and its derivatives is termed as non-linear differential equation. This chapter has been carefully written to provide an easy understanding of linear and non-linear differential equations.

Chapter - Partial Differential Equations: An Integrated Study

Partial differentiation equations comprises of an unknown multivariable and their partial derivatives. First order partial differential equations, second order partial differential equations, fundamental solution, nonlinear partial differential equation, etc. fall within it. All the aspects related to partial differential equations have been carefully analyzed in this chapter.

Chapter - Applications in Various Fields

Differential equations are used in a wide variety of disciplines like from biology, economics, physics, chemistry and engineering. They are used in solving growth and decay problems, temperature problems, dilution problems, electrical circuits, etc. This chapter delves into the varied uses of differential equations for a thorough understanding of the subject.

Alexis Fletcher

1

Understanding Differential Equation

Differential equation can be termed as a relationship between the derivatives of one or more functions. The degree of a differential equation is the highest power of the order derivative and the order of a differential equation refers to the highest number of derivatives in an equation. This is an introductory chapter which will briefly introduce about differential equations.

A differential equation is an equation which contains one or more terms which involve the derivatives of one variable (i.e., dependent variable) with respect to the other variable (i.e., independent variable).

$$dy/dx = f(x)$$

Here “x” is an independent variable and “y” is a dependent variable.

For example,

$$dy/dx = 5x$$

A differential equation that contains derivatives which are either partial derivatives or ordinary derivatives. The derivatives represent a rate of change, and the differential equation describes a relationship between the quantity that is continuously varying and the speed of change. There are a lot of differential equations formulas to the solution of the derivatives.

Differential Equation Types

Differential equations can be divided into several types namely:

- Ordinary Differential Equations,
- Partial Differential Equations,
- Linear Differential Equations,
- Non-linear differential equations,
- Homogeneous Differential Equations,
- Non-homogenous Differential Equations.

Differential Equations Solutions

To find the solution of differential equation, there are two methods to solve differential function:

- Separation of variables,
- Integrating factor.

Separation of the variable is done when the differential equation can be written in the form of $dy/dx = f(y)g(x)$ where f is the function of y only and g is the function of x only. Taking an initial condition We rewrite this problem as $dy/dx = f(y)g(x)$ and then integrate them from both sides.

Integrating factor technique is used when the differential equation is of the form $dy/dx + p(x)y = q(x)$ where p and q are both the functions of x only.

First-order differential equation is of the form $y' + P(x)y = Q(x)$ where p and q are both functions of x and hence called first-order differential equation because it contains functions and the first derivative of y. The higher-order differential equation is an equation that contains derivatives of an unknown function which can be either a partial or ordinary derivative. It can be represented in any order.

Degree of Differential Equation

The degree of the differential equation is the order of the highest order derivative, where the original equation is represented in the form of a polynomial equation in derivatives such as y', y'', y''', and so on.

Suppose $(d^2y/dx^2) + 2(dy/dx) + y = 0$ is a differential equation, so the degree of this equation here is 2. Some more examples are:

- $dy/dx + 1 = 0$, degree is 1.
- $y''' + 3y'' + 6y' - 12 = 0$, degree is 3.

Ordinary Differential Equation

An ordinary differential equation involves function and its derivatives. It contains only one independent variable and one or more of its derivative with respect to the variable.

The order of ordinary differential equations is defined to be the order of the highest derivative that occurs in the equation. The general form of n-th order ODE is given as,

$$F(x, y, y', \ldots, y^n) = 0$$

Applications

Let us see some differential equation applications in real-time:

- Differential equations describe various exponential growths and decays.
- They are also used to describe the change in investment return over time.
- They are used in the field of medicines for modelling cancer growth or the spread of disease in the body.
- Movement of electricity can also be described with the help of it.
- They help economists in finding optimum investment strategies.
- The motion of waves or a pendulum can also be described using these equations.

The various other applications in engineering is: heat conduction analysis, in physics it can be used to understand the motion of waves, pendulums, in chemistry it is used for modelling the chemical reactions, in medical science for monitoring the cancer growth. The ordinary differential equation can be utilized as an application in engineering field like for finding the relationship between various parts of the bridge.

Linear Differential Equations Real World Example

To understand Differential equations, let us consider this simple example. Have you ever thought why a hot cup of coffee cools down when kept under normal conditions? According to Newton, cooling of a hot body is proportional to the temperature difference between its own temperature T and the temperature To of its surrounding. This statement in terms of mathematics can be written as:

$$dT/dt \propto (T - T_0)$$

This is the form of a linear differential equation.

Introducing a proportionality constant k, the above equation can be written as:

$$dT/dt = k(T - T_0)$$

Here, T is the temperature of the body and t is the time:

- T_0 is the temperature of the surrounding.
- dT/dt is the rate of cooling of the body.

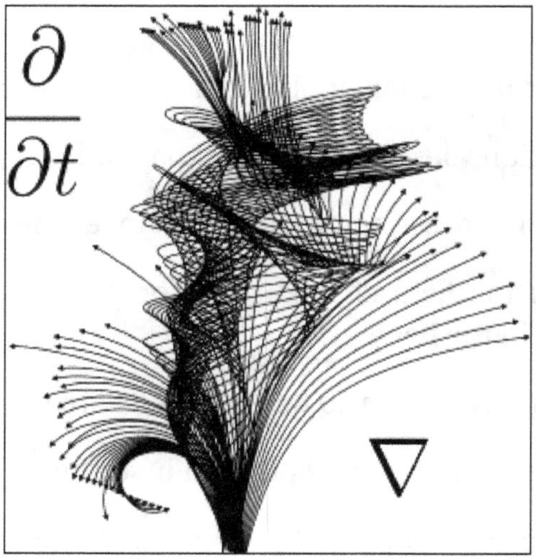

The path of the projectile follows a curve which can be derived from an ordinary differential equation.

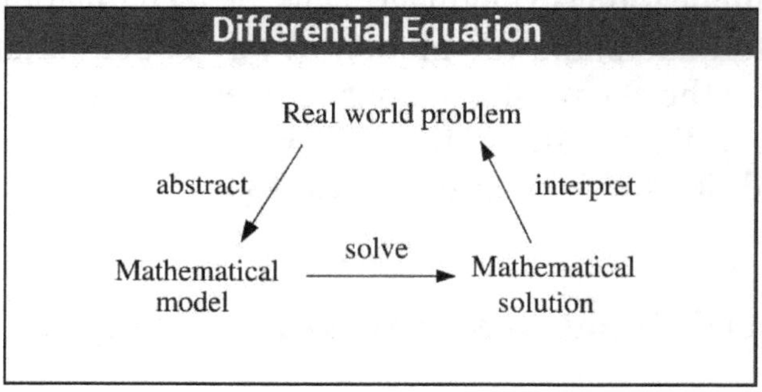

Eg: $dy/dx = 3x$

Here, the differential equation contains a derivative that involves a variable (dependent variable,y) w.r.t another variable(independent variable,x). The types of differential equations are:

- An ordinary differential equation contain one independent variable and its derivatives. It is frequently called as ODE. The general definition of the ordinary differential equation is of the form: Given an F, a function os x and y and derivative of y, we have $F\left(x,y,y'.....y^{\wedge}(n1)\right)=y(n)$ is an explicit ordinary differential equation of order n.

- Partial differential equation that contains one or more independent variable.

Solved problem:

Question: Verify that the function $y = e^{-3x}$ is a solution to the differential equation,

$$\frac{d^2y}{dx^2}+\frac{dy}{dx}-6y=0.$$

Solution: The function given is $y=e^{-3x}$. We differentiate both the sides of the equation with respect to x,

$$\frac{dy}{dx}=-3e^{-3x}$$

Now we again differentiate the above equation with respect to x,

$$\frac{d^2y}{dx^2}=9e^{-3x}$$

We substitute the values of $\frac{dy}{dx},\frac{d^2y}{dx^2}$ and y in the differential equation given in the question, On left hand side we get, $LHS=9e^{-3x}+\left(-3e^{-3x}\right)-6e^{-3x}=9e^{-3x}-9e^{-3x}=0$ (which is equal to RHS).

Therefore the given function is a solution to the given differential equation.

ORDER OF DIFFERENTIAL EQUATIONS

The order of the differential equation is the order of the highest order derivative present in the equation. Here some of the examples for different orders of the differential equation are given:

- $dy/dx=3x+2$, The order of the equation is 1
- $\left(d^2y/dx^2\right)+2\ (dy/dx)+y=0$. The order is 2
- $(dy/dt)+y=kt$. The order is 1.

First Order Differential Equation

You can see in the first example, it is a first-order differential equation which has degree equal to 1. All the linear equations in the form of derivatives are in the first order. It has only the first derivative such as dy/dx, where x and y are the two variables and is represented as,

$$dy/dx=f(x,y)=y'$$

Second Order Differential Equation

The equation which includes second-order derivative is the second-order differential equation. It is represented as,

$$d/dx(dy/dx)=d^2y/dx^2=f''(x)=y''$$

Examples of Differential Equations

It involves a derivative, $\frac{dy}{dx}$:

$$\frac{dy}{dx} = x^2 - 3$$

This will be a general solution (involving K, a constant of integration).

So we proceed as follows:

$$y = \int (x^2 - 3)dx$$

and this gives,

$$y = \frac{x^3}{3} - 3x + K$$

But where did that dy go from the $\frac{dy}{dx}$?

In this example, we appear to be integrating the x part only (on the right), but in fact we have integrated with respect to y as well (on the left). DEs are like that - you need to integrate with respect to two (sometimes more) different variables, one at a time.

We could have written our question only using differentials:

$$dy = (x^2 - 3)dx$$

Now we integrate both sides, the left side with respect to y (that's why we use "dy") and the right side with respect to x (that's why we use "dx"):

$$\int dy = \int (x^2 - 3)dx$$

But this time we have arrived at it considering the dy part more carefully:

$$y = \frac{x^3}{3} - 3x + K$$

On the left hand side, we have integrated $\int dy = \int 1 dy$ to give us y.

We have integrated both sides, but there's a constant of integration on the right side only. What happened to the one on the left? The answer is quite straightforward. We do actually get a constant on both sides, but we can combine them into one constant (K) which we write on the right hand side.

Example:

This example also involves differentials:

$$\theta^2 d\theta = \sin(t+0.2)dt$$

We have:

A function of θ with dθ on the left side,

A function of t with dt on the right side.

To solve this, we would integrate both sides, one at a time, as follows:

$$\int \theta^2\, d\theta = \int \sin(t+0.2)dt$$

$$\frac{\theta^3}{3} = -\cos(t+0.2)+K$$

We have integrated with respect to θ on the left and with respect to t on the right.

Here is the graph of our solution, K=2:

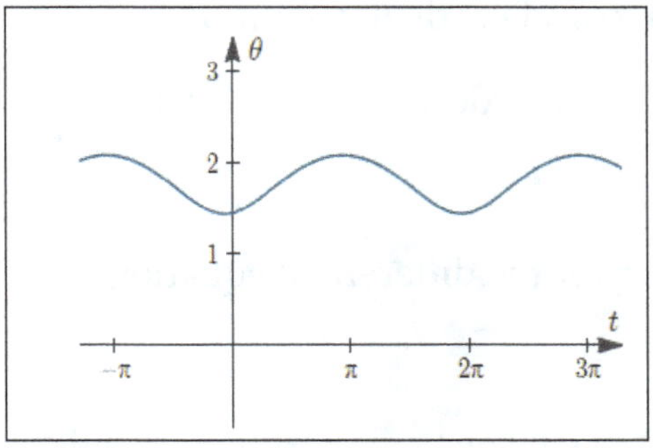

Typical solution graph for the Example DE: $DE: \theta(t) = \sqrt[3]{-3\cos(t+0.2)+6}$.

Solving a Differential Equation

From the above examples, we can see that solving a DE means finding an equation with no derivatives that satisfies the given DE. Solving a differential equation always involves one or more integration steps.

It is important to be able to identify the type of DE we are dealing with before we attempt to solve it.

Example: Order and Degree.

$$\frac{d^2y}{dx^2} + \left(\frac{dy}{dx}\right)^3 - 3x + 2y = 8$$

This DE has order 2 (the highest derivative appearing is the second derivative) and degree 1 (the power of the highest derivative is 1).

$$\left(\frac{dy}{dx}\right)^5 - 2x = 3\sin(x) - \sin(y)$$

This DE has order 1 (the highest derivative appearing is the first derivative) and degree 5 (the power of the highest derivative is 5).

$$(y'')^4 + 2(y')^7 - 5y = 3$$

This DE has order 2 (the highest derivative appearing is the second derivative) and degree 4 (the power of the highest derivative is 4).

General and Particular Solutions

When we first performed integrations, we obtained a general solution (involving a constant, K). We obtained a particular solution by substituting known values for x and y. These known conditions are called boundary conditions (or initial conditions). It is the same concept when solving differential equations - find general solution first, then substitute given numbers to find particular solutions.

Let's see some examples of first order, first degree DEs.

Example:

1. Find the general solution for the differential equation,

$$dy + 7x\,dx = 0$$

Solution: We simply need to subtract 7x dx from both sides, then insert integral signs and integrate:

$$d = -7x\,dx$$

$$\int dy = -\int 7x\,dx$$

$$y = -\frac{7}{2}x^2 + K$$

We are now writing our (simple) example as a differential equation. Earlier, we would have written this example as a basic integral, like this:

$$\frac{dy}{dx} + 7x = 0$$

Then $\frac{dy}{dx} = -7x$ and so $y = -\int 7x\,dx = -\frac{7}{2}x^2 + K$

The answer is the same - the way of writing it, and thinking about it, is subtly different.

$\int dy$ means $\int 1dy$, which gives us the answer y.

We could also have:

$$\int dt = t$$

$$\int d\theta = \theta$$

$$\int da = a$$

and so on.

- Find the particular solution given that y(0)=3.

Solution: We now use the information y(0) = 3 to find K.

The information means that at $x=0, y=3$. We substitute these values into the equation that we found in part (a), to find the particular solution.

$3 = \frac{7}{2}(0)^2 + K$ gives K = 3.

So the particular solution is: $y = -\frac{7}{2}x^2 + 3$, an "n"-shaped parabola.

Here is the graph of the particular solution we just found:

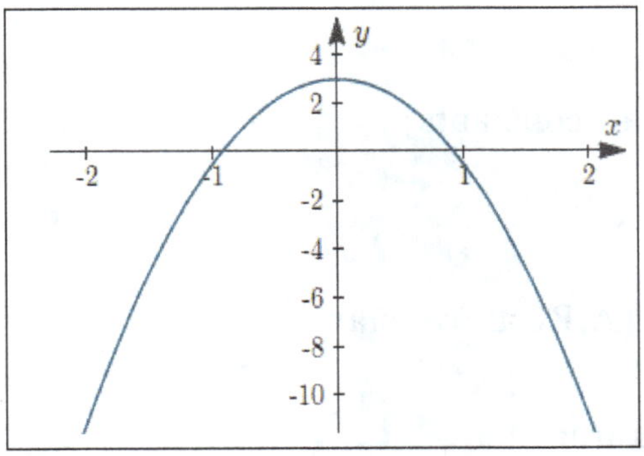

Solution graph: $y = -\frac{7}{2}x^2 + 3$.

Example:

Find the particular solution of,

$$y' = 5$$

given that when $x = 0, y = 2$.

We can write,

$$y' = 5$$

as a differential equation:

$$dy = 5dx$$

Integrating both sides gives:

$$y = 5x + K$$

Applying the boundary conditions: $x = 0, y = 2$, we have $K = 2$ so:

$$y = 5x + 2$$

Example:

Find the particular solution of,

$$y''' = 0$$

given that:

$$y(0) = 3, y'(1) = 4, y''(2) = 6$$

Since $y''' = 0$, when we integrate once we get:

$$y'' = A \quad (A \text{ is a constant})$$

Integrating again gives:

$$y' = Ax + B \quad (A, B \text{ are constants})$$

Once more:

$$y = \frac{Ax^2}{2} + Bx + C \quad (A, B C \text{ are constants})$$

The boundary conditions are:

$$y(0) = 3, y'(1) = 4, y''(2) = 6$$

We need to substitute these values into our expressions for y'' and y' and our general solution,

$$y = \frac{Ax^2}{2} + Bx + C.$$

Now,

$$y(0)=3 \text{ gives } C=3.$$

and

$$y''(2)=6 \text{ gives } A=6$$

(Actually, $y''=6$ for any value of x in this problem since there is no x term)

Finally,

$$y'(1)=4 \text{ gives } B=-2$$

So the particular solution for this question is:

$$y=3x^2-2x+3$$

Checking the solution by differentiating and substituting initial conditions:

$$y'=6x-2$$
$$y'(1)=6(1)-2=4$$
$$y''=6$$
$$y'''=0$$

Example:

After solving the differential equation,

$$\frac{dy}{dx}\ln x-\frac{y}{x}=0$$

$$y=c\ln x$$

Now, if $y=c\ln x$, then $\frac{dy}{dx}=\frac{c}{x}$

So,

$$\begin{aligned} LHS &= \frac{dy}{dx}\ln x-\frac{y}{x} \\ &= \left(\frac{c}{x}\right)\ln x-\frac{(c\ln x)}{x} \\ &= 0 \\ &= RHS \end{aligned}$$

We conclude that we have the correct solution.

Second Order DEs

Example: Show that $y = c_1 \sin 2x + 3\cos 2x$ is a general solution for the differential equation,

$$\frac{d^2y}{dx^2} + 4y = 0$$

We have a second order differential equation and we have been given the general solution. Our job is to show that the solution is correct.

We do this by substituting the answer into the original 2nd order differential equation.

We need to find the second derivative of y:

$$y = c_1 \sin 2x + 3\cos 2x$$

First derivative:

$$\frac{dy}{dx} = 2c_1 \cos 2x - 6\sin 2x$$

Second derivative:

$$\frac{d^2y}{dx^2} = -4c_1 \sin 2x - 12\cos 2x$$

Now for the check step:

$$\begin{aligned} \text{LHS} &= \frac{d^2y}{dx^2} + 4Y \\ &= [-4c_1 \sin 2x - 12\cos 2x] + 4(c_1 \sin 2x + 3\cos 2x) \\ &= 0 \\ &= \text{RHS} \end{aligned}$$

Example:

Show that $\frac{d^2y}{dx^2} = 2\frac{dy}{dx}$ has a solution of $y = c_1 + c_2e^{2x}$

Since $y = c_1 + c_2e^{2x}$, then:

$$\frac{dy}{dx} \quad 2c \;\; e^{2x}$$

and

$$\frac{d^2y}{dx^2} = 4c_2 e^{2x}$$

It is obvious that $\frac{d^2y}{dx^2} = 2\frac{dy}{dx}$.

DERIVATIVE

Derivative is the rate of change of a function with respect to a variable. Derivatives are fundamental to the solution of problems in calculus and differential equations. In general, scientists observe changing systems (dynamical systems) to obtain the rate of change of some variable of interest, incorporate this information into some differential equation, and use integration techniques to obtain a function that can be used to predict the behavior of the original system under diverse conditions.

Geometrically, the derivative of a function can be interpreted as the slope of the graph of the function or, more precisely, as the slope of the tangent line at a point. Its calculation, in fact, derives from the slope formula for a straight line, except that a limiting process must be used for curves. The slope is often expressed as the "rise" over the "run," or, in Cartesian terms, the ratio of the change in y to the change in x. For the straight line shown in the figure, the formula for the slope is $(y_1 - y_0)/(x_1 - x_0)$. Another way to express this formula is $[f(x_0 + h) - f(x_0)]/h$, if h is used for $x_1 - x_0$ and f(x) for y. This change in notation is useful for advancing from the idea of the slope of a line to the more general concept of the derivative of a function.

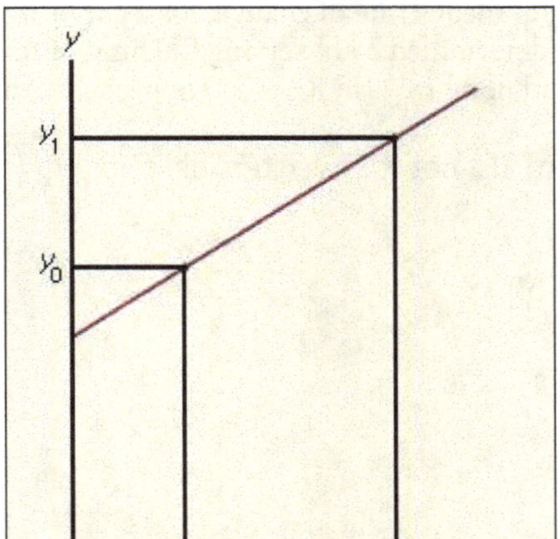

Two points, such as (x_0, y_0) and (x_1, y_1), determine the slope of a straight line.

For a curve, this ratio depends on where the points are chosen, reflecting the fact that curves do not have a constant slope. To find the slope at a desired point, the choice of the second point needed to calculate the ratio represents a difficulty because, in general, the ratio will represent only an average slope between the points, rather than the actual slope at either point. To get around this difficulty, a limiting process is used whereby the second point is not fixed but specified by a variable, as h in the ratio for the straight line above. Finding the limit in this case is a process of finding a number that the ratio approaches as h approaches 0, so that the limiting ratio will represent the actual slope at the given point. Some manipulations must be done on the quotient $[f(x_0 + h) - f(x_0)]/h$ so that it can be rewritten in a form in which the limit as h approaches 0 can be seen more directly. Consider, for example, the parabola given by x^2. In finding the derivative of x^2 when x is 2, the quotient is $[(2 + h)^2 - 2^2]/h$. By expanding the numerator, the quotient becomes $(4 + 4h + h^2 - 4)/h = (4h + h^2)/h$. Both numerator and denominator still approach 0, but if h is not actually zero but only very close to it, then h can be divided out, giving $4 + h$, which is easily seen to approach 4 as h approaches 0.

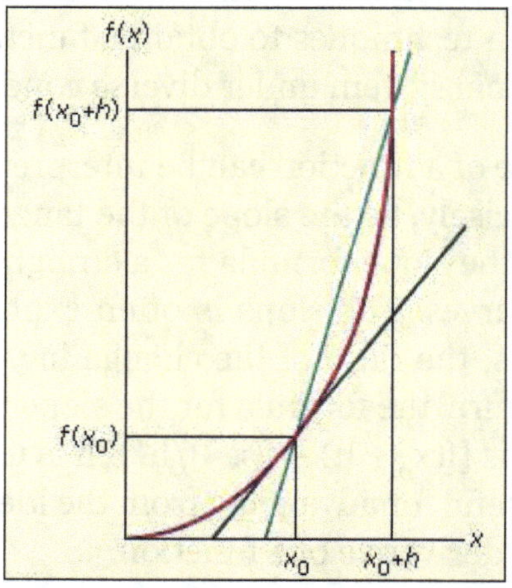

The slope, or instantaneous rate of change, for a curve at a particular point $(x_0, f(x_0))$ can be determined by observing the limit of the average rate of change as a second point $(x_0 + h, f(x_0 + h))$ approaches the original point.

To sum up, the derivative of f(x) at x_0, written as $f'(x_0)$, $(df/dx)(x_0)$, or $Df(x_0)$, is defined as,

$$\lim_{h\to 0}\left[f(x_0+h)-f(x_0)\right]/h$$

if this limit exists.

FRACTIONAL DIFFERENTIAL EQUATIONS

Fractional calculus is the theory of integrals and derivatives of arbitrary real (and even complex) order and was first suggested in works by mathematicians such as Leibniz,

L'Hôpital, Abel, Liouville, Riemann, etc. The importance of fractional derivatives for modeling phenomena in different branches of science and engineering is due to their nonlocality nature, an intrinsic property of many complex systems. Unlike the derivative of integer order, fractional derivatives do not take into account only local characteristics of the dynamics but considers the global evolution of the system; for that reason, when dealing with certain phenomena, they provide more accurate models of real-world behavior than standard derivatives.

To illustrate this fact, we will retrieve an example from equation $\sigma(t)=\eta\frac{d}{dt}\varepsilon(t)$. Recall the relationship between stress σ(t) and strain ε(t) in a material under the influence of external forces:

$$\sigma(t)=\eta\frac{d}{dt}\varepsilon(t)$$

is the Newton's law for a viscous liquid, with η the viscosity of the material,

$$\sigma(t)=E\varepsilon(t)$$

is Hooke's law for an elastic solid, with E the modulus of elasticity. We can rewrite both equation $\sigma(t)=\eta\frac{d}{dt}\varepsilon(t)$ and $\sigma(t)=E\varepsilon(t)$ as,

$$\sigma(t)=v\frac{d^{\alpha}}{dt^{\alpha}}\varepsilon(t)$$

with $\alpha=0$ for elastic solids and $\alpha=1$ for a viscous liquid. But, in practice, there exist viscoelastic materials that have a behavior intermediate between an elastic solid and a viscous liquid, and it may be convenient to give sense to the operator $\frac{d^{\alpha}}{dt^{\alpha}}$ if $0<\alpha<1$. There exist various definitions of fractional derivatives. All of them involve integral operators with different regularity properties, and some of them have singular kernels.

The classical Cauchy formula for the n -fold iterated integral, with $n\in\mathbb{N}$ is,

$$_0I_t^n[f](t)=\frac{1}{(n-1)!}\int_0^t(t-s)^{n-1}f(s)ds.$$

Recalling that gamma function verifies $n\Gamma(n)=n!$, an immediate generalization of this formula for a real order α is,

$$_0I_t^{\alpha}[f](t)=\frac{1}{\Gamma(\alpha)}\int_0^t(t-s)^{\alpha-1}f(s)ds,$$

known as Riemann-Liouville fractional integral operator of order α (the term "fractional" is misleading but has a historical origin). From this, the Riemann-Liouville fractional derivative of order α, with $n-1<\alpha<n$ is defined as,

$$ {}^{RL}_{0}D^{\alpha}_{t}[f](t)=\frac{1}{\Gamma(n-\alpha)}\frac{d^n}{dt^n}\int_0^t (t-s)^{n-\alpha-1} f(s)ds. $$

while the Caputo fractional derivative of order α is defined as,

$$ {}^{C}_{0}D^{\alpha}_{t}[f](t)=\frac{1}{\Gamma(n-\alpha)}\int_0^t (t-s)^{n-1-\alpha}\frac{d^n}{ds^n}[f(s)]ds. $$

Both equations ${}^{RL}_{0}D^{\alpha}_{t}[f](t)=\frac{1}{\Gamma(n-\alpha)}\frac{d^n}{dt^n}\int_0^t (t-s)^{n-\alpha-1} f(s)ds$ and ${}^{C}_{0}D^{\alpha}_{t}[f](t)=\frac{1}{\Gamma(n-\alpha)}\int_0^t (t-s)^{n-1-\alpha}\frac{d^n}{ds^n}[f(s)]ds$ define left inverse operators for the integral operator of Riemann-Liouville of order α and are associated to the idea that "deriving α times may be thought as integrating n−α times and deriving n times." Of course these definitions aren't equivalent: clearly the domains of the operators ${}^{RL}_{0}D^{\alpha}_{t}[.]$ and $D\ [.]$ are different; because of the different hypothesis about f, we need to impose to guarantee their existence. Besides that, with the appropriate conditions for f,

$$ {}^{C}_{0}D^{\alpha}_{t}[f](t)={}^{RL}_{0}D^{\alpha}_{t}[f](t)-\sum_{k=0}^{\infty}\frac{t^{k-\alpha}}{\Gamma(k-\alpha+1)}f^{(k)}(0^+). $$

More recently, the Caputo-Fabrizio fractional derivative of order α, with $\alpha\in[0,1)$, was defined as,

$$ {}^{CF}_{0}D^{\alpha}_{t}[f](t)=\frac{M(\alpha)}{1-\alpha}\int_0^t f'(s)e^{-\frac{\alpha(t-s)}{1-\alpha}}ds, $$

for M (α) is a normalizing factor verifying $M(0)=M(1)=1$.

Let us point out that, both in equation ${}^{C}_{0}D^{\alpha}_{t}[f](t)=\frac{1}{\Gamma(n-\alpha)}\int_0^t (t-s)^{n-1-\alpha}\frac{d^n}{ds^n}[f(s)]ds$ and ${}^{CF}_{0}D^{\alpha}_{t}[f](t)=\frac{M(\alpha)}{1-\alpha}\int_0^t f'(s)e^{-\frac{\alpha(t-s)}{1-\alpha}}ds$, the lower limit in the integral could be changed by any value $a\in[-\infty,t)$, i.e.

$$ {}^{C}_{0}D^{\alpha}_{t}[f](t)=\frac{1}{\Gamma(n-\alpha)}\int_a^t (t-s)^{n-1-\alpha}\frac{d^n}{ds^n}[f(s)]ds $$

and

$$ {}^{CF}_{a}D_t^{\alpha}[f](t) = \frac{M(\alpha)}{1-\alpha}\int_0^t f'(s)e^{-\frac{\alpha(t-s)}{1-\alpha}}ds. $$

In equation ${}^{C}_{0}D_t^{\alpha}[f](t) = \frac{1}{\Gamma(n-\alpha)}\int_0^t (t-s)^{n-1-\alpha}\frac{d^n}{ds^n}[f(s)]ds$ the researchers prove that the operator defined in equation ${}^{CF}_{0}D_t^{\alpha}[f](t) = \frac{M(\alpha)}{1-\alpha}\int_0^t f'(s)e^{-\frac{\alpha(t-s)}{1-\alpha}}ds$, verifies the following (convenient) properties:

${}^{CF}_{0}D_t^{\alpha}[k] = 0$, for any constant k.

$$ \lim_{\alpha \to 1} {}^{CF}_{0}D_t^{\alpha}[f](t) = \frac{df}{dt}. $$

$$ \lim_{\alpha \to 0} {}^{CF}_{0}D_t^{\alpha}[f](t) = f(t) - f(0). $$

Caputo-Fabrizio definition was then generalized by Atangana and Baleanu, who gave the following definition of the Atangana-Baleanu fractional derivative in Riemann-Liouville sense:

$$ {}^{ABR}_{0}D_t^{\alpha}[f](t) = \frac{M(\alpha)}{1-\alpha}\frac{d}{dt}\int_0^t f(s)E_{\alpha}\left(-\frac{\alpha(t-s)^{\alpha}}{1-\alpha}\right)ds $$

and the Atangana-Baleanu fractional derivative in Caputo sense,

$$ {}^{ABC}_{0}D_t^{\alpha}[f](t) = \frac{M(\alpha)}{1-\alpha}\int_0^t f'(s)E_{\alpha}\left(-\frac{\alpha(t-s)^{\alpha}}{1-\alpha}\right)ds $$

replacing the exponential by $E_{\alpha}(z) = \sum_{k=0}^{\infty}\frac{z^k}{\Gamma(\alpha k+1)}$, the generalized Mittag-Leffler function.

Other types of fractional derivatives are Grünwald-Letnikov's, Hadamard's, Weyl's, etc. In every definition it is clear that fractional derivative operators are not local, since they need the information of f in a whole interval of integration. When defined with 0 as lower limit of integration, as we did, function f is usually assumed to be causal (i.e., $f(t) \equiv 0$ for $t < 0$), but this limit can also be changed.

Caputo and Fabrizio proposed the following terms to recognize if an integral operator merits to be called a fractional derivative:

- The fractional derivative must be a linear operator.

- The fractional derivative of an analytic function must be analytic.
- If the order of the derivative is a positive integer, the derivative must be the classical one.
- If the order is null, the original function must be recovered.
- For $n \in \mathbb{N}, 0 < \alpha < 1: D_t^\alpha\left[D_t^n[f]\right](t) = D_t^n\left[D_t^\alpha[f]\right](t)$.
- $D_t^\alpha[f](t)$ must depend on the past history of f.

Many applications of fractional calculus have been reported in areas as diverse as diffusion problems, hydraulics, potential theory, control theory, electrochemistry, viscoelasticity, and nanotechnology, among others.

Theoretical results concerning existence and uniqueness of solutions to fractional differential equations have been also developed.

The researchers state conditions to guarantee the existence and uniqueness of solution to problems like,

$$\begin{cases} {}_0^C D_t^\alpha[f](t) = F(t, f(t)) t \in (0, T), T < \infty \\ \text{initial or boundary conditions} \end{cases}$$

or

$$\begin{cases} {}_0^{RL} D_t^\alpha[f](t) = F(t, f(t)) t \in (0, T), T < \infty \\ \text{initial or boundary conditions} \end{cases}$$

for $0 < \alpha < 2$. After rewriting the equation as an integral equation with a kernel whose norm is bounded in a proper Banach space, they use generalizations of the fixed-point theorem. The function F, besides being continuous, must satisfy certain conditions that substitute the classical Lipschitz's one.

Similar results are stated in equation below for a coupled system of fractional differential equations involving Riemann-Liouville derivatives.

Analytical calculus of fractional operators is, in general, difficult. In existence and uniqueness, for the solution to a simple case,

$$\begin{cases} {}_0^{CF} D_t^\alpha[f](t) = \beta f(t) = g(t) \\ f(0) = 0 \end{cases}$$

are proved, and explicit formulae are presented when g is continuous, causal, and null at the origin. The case of Caputo derivative is also considered. In all cases, the computation of the primitive of the data function is required.

Numerical methods have also been proposed to obtain approximate solution to fractional differential equations. In some numerical approximations to solutions to different fractional differential equations are presented and experimentally verified on various examples, and complete surveys on numerical methods are offered. A local fractional natural homotopy perturbation method is proposed to found the solution to partial fractional differential equations as a series.

A method based on a semi-discrete finite difference approximation in time, and Galerkin finite element method in space to solve fractional partial differential equations arising in neuronal dynamics. A new numerical approximation of Atangana-Baleanu integral, as the summation of the average of the given function and its fractional integral in Riemann-Liouville sense, is proposed. Semi-discrete finite element methods are introduced to solve diffusion equations, and implicit numerical algorithms for the case of spatial and temporal fractional derivatives appeared. A high-speed numerical scheme for fractional differentiation and fractional integration is proposed. A new numerical method to solve partial differential equations involving Caputo derivatives of fractional variable order is obtained in terms of standard (integer order) derivatives.

A discrete form is proposed for solving time fractional convection-diffusion equation. The Laplace transform is used to solve fractional differential equations.

Some Mathematical Models with Fractional Derivatives

The researchers review the evolution of the general fractional equation:

$$\frac{\partial^{\beta} u}{\partial t^{\beta}} = a \frac{\partial^2 u}{\partial x^2}$$

for $a > 0, 0 < \beta \leq 2$, where $x \in S \subset \mathbb{R}$ and $t \in \mathbb{R} > 0$ denote the space and time variables. This equation is obtained from the classical D'Alembert wave equation by replacing the second-order time derivative with the Caputo fractional derivative of order $\beta \in (0,2]$. The researchers show that, for $1 < \beta < 2$, the behavior of the fundamental solutions turns out to be intermediate between diffusion (for a viscous fluid) and wave propagation (for an elastic solid), thus justifying the attribute of fractional diffusive waves.

Fractional wave is solved, for $0 < \beta < 1$ and special initial conditions, by the method of separation of variables.

The researchers study the particular linear fractional Klein-Gordon equation:

$$\begin{cases} \dfrac{\partial^{\alpha} u}{\partial t^{\alpha}} - \dfrac{\partial^2 u}{\partial x^2} + u = 6x^3 t + (x^3 - 6x)t^3 \quad t > 0, x \in \mathbb{R} \\ u(x,0) = 0 \\ u_t(x,0) = 0 \end{cases}$$

considering the fractional Caputo derivative with $1 < \alpha \leq 2$. They achieve a numerical solution by variational iteration method and multivariate Padé approximation.

The following mathematical model, using Fick's law of diffusion, is developed to study the effect of fractional advection diffusion equation (cross flow) for the calcium profile, considering the Caputo fractional derivative $(0 < \alpha \leq 1)$:

$$\begin{cases} \dfrac{\partial^\alpha C}{\partial t^\alpha} = D\dfrac{\partial^2 C}{\partial x^2} & x \geq 0, t \geq 0 \\ C(x,0) = C_0 & \\ \lim\limits_{x \to +\infty} C(x,t) = C_\infty & \end{cases}$$

Here, the calcium concentration $C(x,t)$ varies in time and space, D is a diffusion constant, and C_∞ is calcium concentration at infinity and is assumed that, at initial state of time and at a long distance, calcium concentration vanishes or becomes zero. The researchers note that the physical parameter α characterizes the cytosolic calcium ion in astrocytes.

The researchers explain that arteries, like other soft tissues, exhibit viscoelastic behavior and part of the mechanical energy transferred to them is dissipative (viscosity) and the other part is stored in a reversible form (elasticity). They modified the standard model by a fractional-order one and test it in human arterial segments. They conclude that fractional derivatives, in Riemann-Liouville sense, are a good alternative to model arterial viscosity.

The generalized Voigt model consists of a spring in parallel with two springpots (a neologism for a model that is between a spring—purely elastic—and a dashpot, purely viscous) of fractional orders α and β. The governing fractional-order differential equation is,

$$\sigma(t) = E_0\varepsilon(t) + \eta_1 \frac{\partial^\alpha \varepsilon(t)}{\partial t^\alpha} + \eta_2 \frac{\partial^\beta \varepsilon(t)}{\partial x^\beta}$$

where E_0 is the elastic constant for a spring and η_1 and η_2 represent the viscosities of two spring pots in parallel with the spring.

An accurate and efficient numerical method for the fractional-order standard model described by equation above with $\alpha=\beta$, combining a fast convolution method with the spectral element discretization based on a general Jacobi polynomial basis that can be used to generate 3D polymorphic high-order elements. In that way they model complicated arterial geometries, such as patient-specific aneurysms, and apply it to 3D fluid-structure interaction simulations.

The researchers use fractional derivatives to model the magnetic hysteresis, a phenomenon where the "memory" of the ferromagnetic material is crucial. They use a nonlinear model for the constitutive law of an isotropic ferromagnetic material:

$$\lambda H(t)=\theta_c\left(M^2(t)+1\right)M(t)-\theta(t)M(t)-C_0 D_t^\alpha[M](t)$$

for $\lambda > 0, H(t)$ is the magnetic excitation field, M(t) is the magnetization vector, θ(t) is the temperature, θ_c is the Curie temperature below which the hysteresis is observed, and C_0 is the tensor with the constitutive properties of the magnetic material. They compare the resulting behavior when D_t^α is the Caputo fractional derivative with the one that results when the derivative is the Caputo-Fabrizio one. By numerical simulations they obtain examples of the classical hysteresis cycles and conclude that Caputo derivative expresses a stronger memory than the Caputo-Fabrizio operator.

These are just a few examples of the huge variety of problems that can be modeled by means of fractional differential equations. The nonlocality of the associated operators is the key to the success in the description of these phenomena.

Some Simple Systems Exhibiting Chaos

Chaos theory is also an area where fractional derivatives play an important role.

The researchers studied a system based on the classical Lorenz one, but described by the Atangana-Baleanu fractional derivative (in the Caputo sense) with $0<\alpha<1$:

$$\begin{cases} {}^{ABC}_{\;\;0}D_t^\alpha[x](t)=\sigma y(t)-\alpha x(t) \\ {}^{ABC}_{\;\;0}D_t^\alpha[y](t)=\rho x(t)-x(t)z(t)-y(t) \\ {}^{ABC}_{\;\;0}D_t^\alpha[z](t)=x(t)y(t)-\beta z(t) \end{cases}$$

Under certain assumptions on the physical problem, they proved existence of solutions, and, by means of an iterative algorithm, numerical evidence of chaos is shown when $0.25<\alpha<0.3$ and $0.4<\alpha<0.5$ for the usual set of parameters $\sigma=10, \rho=\frac{8}{3}$, and $\beta=28$.

A three-dimensional fractional-order dynamical system for cancer growth is proposed replacing the standard derivatives in the evolution equations:

$$\begin{cases} \dot{x}_1(t)=x_1(t)-Ax_1(t)x_2(t)-Bx_1(t)x_3(t) \\ \dot{x}_2(t)=Cx_2(t)\left(1-x_2(t)\right)-Dx_1(t)x_2(t) \\ \dot{x}_3(t)=E\dfrac{x_1(t)x_3(t)}{x_1(t)+F}-Gx_1(t)x_2(t)-Hx_3(t) \end{cases}$$

by the Caputo-Fabrizio and the Atangana-Baleanu (Caputo sense) derivatives. The system parameters are related to the rate of change in the population of the different cells: healthy and tumor ones. The researchers prove that the system has a unique solution and show that the system exhibits chaos for a proper choice of the parameters values and initial conditions.

A fractional Lorenz system is studied considering the generalized Caputo derivative, defined, for $0<\alpha<1, \rho>0$, as,

$$ {}^{GC}_{\ 0}D^{\alpha}_{t}[f](t)=\frac{\rho^{\alpha}}{\sigma(1-\alpha)}\int_{0}^{t}\frac{f'(s)}{(t^{\rho}-s^{\rho})^{\alpha}}ds $$

A detailed analysis of the stability of the system is performed. Adomian method is used to find semi-analytical solution to the fractional nonlinear equations. Chaotic behavior and strange attractors are numerically found for some values of α and ρ.

A Numerical Approximation Scheme to Solve Linear Fractional Differential Equations

After having exemplified several applications of fractional differential equations to different real-world problems, including chaotic ones, we will show a method that we have developed to obtain numerical solutions to linear fractional initial value and boundary value problems modeled with Caputo or Caputo-Fabrizio derivatives. The idea of the approximation scheme is to transform the derivatives into integral operators acting on the Fourier transform and to perform a wavelet decomposition of the data. The wavelet coefficients of the unknown are then recovered from a linear system of algebraic equations, and the solution is built up from its coefficients. The properties of the chosen wavelet basis guarantee numerical stability and efficiency of the approximation scheme. In the case of singular kernel, this procedure enables us to handle the singularity.

We note that choosing $a=-\infty$ in definition of equation ${}_{0}I^{n}_{t}[f](t)=\frac{1}{(n-1)!}\int_{0}^{t}(t-s)^{n-1}f(s)ds$ or ${}_{0}I^{\alpha}_{t}[f](t)=\frac{1}{\Gamma(\alpha)}\int_{0}^{t}(t-s)^{\alpha-1}f(s)ds$, and being $f\in H^{1}(-\infty,b)$, the Sobolev space of functions with (weak) first derivative in $L^{2}(-\infty,b)$, both derivatives can be expressed as a convolution.

For the Caputo-Fabrizio fractional derivative of order $0<\alpha<1$, changing variables in equation ${}^{CF}_{\ 0}D^{\alpha}_{t}[f](t)=\frac{M(\alpha)}{1-\alpha}\int_{0}^{t}f'(s)e^{-\frac{\alpha(t-s)}{1-\alpha}}ds$, we have,

$$ {}^{CF}_{-\infty}D^{\alpha}_{t}[f](t)=\frac{M(\alpha)}{1-\alpha}\int_{0}^{\infty}f'(t-s)e^{-\frac{\alpha}{1-\alpha}s}ds=\frac{M(\alpha)}{1-\alpha}(f'*k)(t) $$

where k is a causal function, $k(t)=e^{-\frac{\alpha t}{1-\alpha}}$ for $t \geq 0$, and $k(t)=0$ for $t<0$.

Consequently, since $\hat{k}(\omega)=\frac{1-\alpha}{\alpha+i\omega(1-\alpha)}$,

$$ {}_{-\infty}^{CF}D_t^{\alpha}[f](t)=\frac{M(\alpha)}{2\pi(1-\alpha)}\int_R \hat{f}'(\omega)\hat{k}(\omega)e^{i\omega t}d\omega. $$

Using the properties of the Fourier transform, we can rewrite the last equality:

$$ {}_{-\infty}^{CF}D_t^{\alpha}[f](t)=\frac{M(\alpha)}{1-\alpha}\int_R \hat{f}(\omega)h(\omega)e^{i\omega t}d\omega $$

where $h(\omega)=\frac{i\omega}{2\pi}\hat{k}(\omega)$.

Meanwhile, in the Caputo case, we have,

$$ {}_{-\infty}^{CF}D_t^{\alpha}[f](t)=\frac{1}{\Gamma(1-\alpha)}\int_{-\infty}^{t}\frac{f'(s)}{(t-s)^{\alpha}}ds=\frac{1}{2\pi\Gamma(1-\alpha)}\int_R \hat{f}(\omega)\hat{k}(\omega)e^{i\omega t}d\omega $$

where $k(t)=\frac{1}{t^{\alpha}}$ and $\hat{k}(\omega)=\frac{\Gamma(1-\alpha)}{(i\omega)^{1-\alpha}}$.

Initial Value Problem

Let us consider the initial value problem (IVP):

$$ \begin{cases} {}_{0}^{CF}D_t^{\alpha}[f](t)+\sigma_0 f(t)+\sigma_1 f'(t)=g(t) \\ f(0)=0 \end{cases} $$

We look for f satisfying equation above, where g is a causal and smooth function with g(0)=0. Other situations where the initial condition is not null can also be faced adapting the following scheme. We will consider that the fractional derivative is the Caputo-Fabrizio one; the Caputo case can be solved similarly.

First we choose a wavelet basis with special properties: well localized in both time and frequency domain, smooth, band limited and infinitely oscillating with fast decay. The mother wavelet $\psi \in S$ (the Schwartz space) and its Fourier transform satisfy $\mathrm{supp}|\hat{\psi}_{jk}|=\Omega_j$ where $\Omega_j=\{\omega: 2^j(\pi-\beta)\leq|\omega|\leq 2^j(\pi+\beta)\}$, with $0<\beta\leq\frac{\pi}{3}$.

The family $\left\{\psi_{jk}/\psi_{jk}=2^{\frac{j}{2}}\psi(2^j t-k), j,k\in\mathbb{Z}\right\}$ is an orthonormal basis (BON) of $L^2(\mathbb{R})$ associated to a multiresolution analysis (MRA). We denote, by $W_j=\mathrm{span}\{\psi_{jk}, j,k\in\mathbb{Z}\}$

and $V_J = \oplus_j < JW_j$, the wavelet and scale subspaces, respectively, and decompose the space $L^2(\mathbb{R}) = \oplus_{j\in\mathbb{Z}} W_j = \oplus_{j>n} W_j + V_n, n \in \mathbb{Z}$. We have also a scale function $\varphi \in V_0$ so that $\{\varphi(t-k), k \in \mathbb{Z}\}$ is a BON of V_0.

The sets $\Omega_{j-1}, \Omega_j, \Omega_{j+1}$ have little overlap, and W_j is nearly a basis for the set of functions whose Fourier transform has support in Ω_j. This property of the basis will be crucial in the procedure.

Now we decompose the data g as,

$$g(t) = \sum_{n\in N} \langle g, \varphi_{Jn} \rangle \varphi_{Jn}(t) + \sum_{j>J} \sum_{k\in\mathbb{Z}} \langle g, \psi_{jk} \rangle \psi_{jk}(t)$$

where the first and second terms are the projections of g in V_J and W_j, respectively.

The properties of localization of the wavelets guarantee absolute convergence in each W_j.

Now we choose the levels where the energy of g is concentrated, $J_{\min}, J_{\max} \in \mathbb{Z}$ so that,

$$g(t) = \sum_{J_{\min}}^{J_{\max}} g_j(t) + r(t) = \sum_{J_{\min}}^{J_{\max}} \sum_{k\in\mathbb{Z}} \langle g, \psi_{jk} \rangle \psi_{jk}(t) + r(t), \|r\|_2 \le \varepsilon \|g\|_2$$

for small ε, and truncate the component in each level so that the following approximation of g_j arises:

$$\tilde{g}_j(t) = \sum_{k\in k_j} c_{jk} \psi_{jk}(t), c_{jk} = g, \psi_{jk}.$$

Afterwards we obtain the fractional Caputo-Fabrizio derivatives of the wavelet basis by means of equation ${}^{CF}_{-\infty}D^{\alpha}_t[f](t) = \frac{M(\alpha)}{1-\alpha} \int_R \hat{f}(\omega) h(\omega) e^{i\omega t} d\omega$:

$$v_{jk}(t) = {}^{CF}_{-\infty}D^{\alpha}_t \left[\psi_{jk}\right](t) = \frac{M(\alpha)}{1-\alpha} \int_{\Omega_j} \hat{\psi}_{jk}(\omega) h(\omega) e^{i\omega t} d\omega$$

(recall $\sup|\hat{\psi}_{jk}| = \Omega_j$).

Let us consider for a moment that in equation $\begin{cases} {}^{CF}_{0}D^{\alpha}_t[f](t) + \sigma_0 f(t) + \sigma_1 f'(t) = g(t) \\ f(0) = 0 \end{cases}$ we have ${}^{CF}_{-\infty}D^{\alpha}_t$, i.e., $a = -\infty$. Since $\mathrm{supp}|v_{jk}| \subset \Omega_j, v_{jk} \in W_{j-1} \cup W_j \cup W_{j+1}$, but, from the properties of the chosen basis, we can consider $v_{jk} \in W_j$. This fact enables us to work on each level j separately.

Then, since the unknown f can be expressed as $f(t)=\sum_{j\in\mathbb{Z}} f_j(t)$, where $f_j(t)=\sum_{k\in\mathbb{Z}} b_{jk}\psi_{jk}(t)$ and $b_{jk}=\langle f,\phi_{jk}\rangle$, we have,

$$f(t)\approx\sum_{J_{min}}^{J_{max}} f_j(t)\approx\sum_{J_{min}}^{J_{max}}\sum_{k\in K_j} b_{jk}\psi_{jk}(t).$$

Finally, we replace this last expression in equation,

$$\begin{cases} {}_{0}^{CF}D_t^{\alpha}[f](t)+\sigma_0 f(t)+\sigma_1 f'(t)=g(t) \\ f(0)=0 \end{cases}$$

where for simplicity we will first consider $\sigma_1=0$ and look for the wavelet coefficients b_{jk} that satisfy, for each $\in[J_{min},J_{max}]$:

$$\sum_{k\in K_j} b_{jk}v_{jk}(t)+\sigma_0\sum_{k\in K_j} b_{jk}\psi_{jk}(t)=\sum_{k'\in K_j} c_{jk'}\psi_{jk'}(t)$$

or,

$$\sum_{k\in K_j} b_{jk}\langle v_{jk},\psi_{jm}\rangle+\sigma_0\sum_{k\in K_j} b_{jk}\langle\psi_{jk},\psi_{jm}\rangle=\sum_{k'\in K_j} c_{jk'}\langle\psi_{jk'}\psi_{jm}\rangle$$

that, in matrix form, results in:

$$\mathcal{M}^j\mathbf{b}^j=\mathbf{c}^j$$

where $\mathbf{b}^j=(b_{jk})_{k\in k_j}$, $\mathbf{c}^j=(c_{jk}')_{k'\in k_j}$, $\mathcal{M}^j\in\mathbb{R}^{K_j\times K_j}$ and $(\mathcal{M}^j)_{kl}=\langle v_{kl},\psi_{kl}\rangle+\sigma_0\langle\psi_{kj},\psi_{kl}\rangle$. From the properties of the wavelet basis, and those of v_{jk}, it results that $\mathcal{M}^j$ is a diagonal dominant matrix and, consequently, the vector of coefficients $\mathbf{b}^j$ can be computed in a stable and accurate way. The solution f can be obtained from equation,

$$f(t)\approx\sum_{J_{min}}^{J_{max}} f_j(t)\approx\sum_{J_{min}}^{J_{max}}\sum_{k\in K_j} b_{jk}\psi_{jk}(t).$$

Moreover, it can be shown that f(0)=0 . To correct the effect of having considered ${}_{-\infty}^{CF}D_t^{\alpha}$, instead of ${}_{0}^{CF}D_t^{\alpha}$, we set $\tilde{f}=f.\chi_{[0,T]}$, where $\chi_{[0,T]}$ is the characteristic function of the interval $[0,T]$. Finally, $\tilde{f}$ is an approximate solution to equation,

$$\begin{cases} {}_{0}^{CF}D_t^{\alpha}[f](t)+\sigma_0 f(t)+\sigma_1 f'(t)=g(t) \\ f(0)=0 \end{cases}.$$

The error introduced in the approximation can be controlled and reduced: a more accurate truncated projection of the data into the wavelet subspaces can be considered, and the elements of the matrix can be computed with good precision since they can be expressed as integrals over compact subsets; finally, the matrix of the resulting linear system is a diagonal dominant matrix, and the solution can be computed accurately. In summary, the good properties of the basis and the operator guarantee that the resulting approximation scheme is efficient and numerically stable and no additional conditions need to be imposed.

Example:

We illustrate the performance of the proposed approximation scheme by solving the IVP described by equation $\begin{cases} {}^{CF}_{0}D^{\alpha}_{t}[f](t)+\sigma_0 f(t)+\sigma_1 f'(t)=g(t) \\ f(0)=0 \end{cases}$ for $\sigma_0 = 0.9$, $\sigma_1 = 0$, and $\alpha = 0.5$ and g a causal function defined as $g(t)=v(t)\sin(2\pi t)\cos(0.5\pi t)$, where v is a smooth window in the interval. Wavelet analysis indicates that the energy of the data g is concentrated in the subspaces W_0 and W_1; thus, we consider levels $-1 \le j \le 2$ for the reconstruction.

For this case, being $\sigma_1 = 0$, $\sigma_0 \ne -\frac{1}{1-\alpha}$, $g \in C(0,\infty)$, and $g(0)=0$, there exists a formula for the "exact" solution to equation $\begin{cases} {}^{CF}_{0}D^{\alpha}_{t}[f](t)+\sigma_0 f(t)+\sigma_1 f'(t)=g(t) \\ f(0)=0 \end{cases}$. The approximate solution $\tilde{f}$ to the IVP is plotted (in green) in figure, together with the exact solution (in blue).

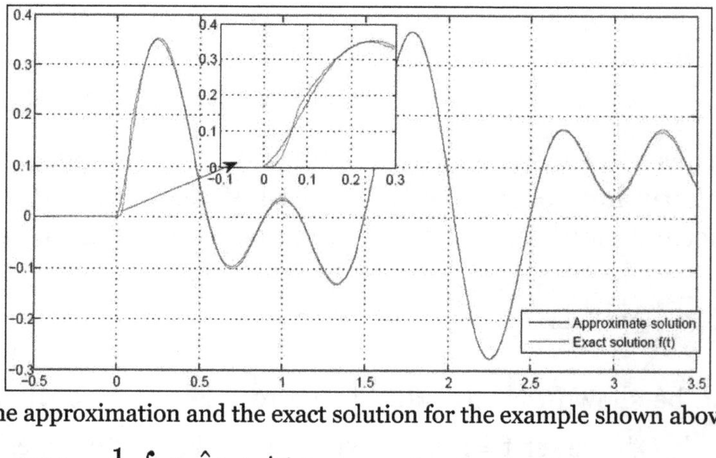

The approximation and the exact solution for the example shown above.

If $\sigma_1 \ne 0$, since $f'(t)=\frac{1}{2\pi}\int_{\mathbb{R}} i\omega\hat{f}(\omega)e^{i\omega t}d\omega$, we obtain similar equations for the coefficients on each level:

$$\sum_{k\in K_j} b_{jk}\langle v_{jk},\psi_{jm}\rangle+\sigma_0\sum_{k\in K_j} b_{jk}\langle \psi_{jk},\psi_{jm}\rangle+\sigma_1\sum_{k\in K_j} b_{jk}\langle i\omega\psi_{jk'}\psi_{jm}\rangle =$$
$$\sum_{k'\in K_j} c_{jk'}\langle \psi_{jk'},\psi_{jm}\rangle$$

Once more the matrix of the resulting linear system is diagonal dominant, and the system can be solved efficiently.

The following example shows the performance of the method for $\sigma_1 = 0.3$.

Example:

Now we consider IVP described by equation $\begin{cases} {}^{CF}_{0}D^{\alpha}_{t}[f](t) + \sigma_0 f(t) + \sigma_1 f'(t) = g(t) \\ f(0) = 0 \end{cases}$ for $ó_1 = 0.9, ó_1 = 0.3, \ \alpha = 0.5,$ and $f(t) = t^2 v(t)\left(2\sin(2.5\pi t) - 0.5\cos(6\partial t)\right),$ with v as a smooth window over interval [0, 7]. Since $\sigma_1 \neq 0$ and we have no formula to test the performance of the approximation, for this example we will set f, calculate g from equation,

$$\begin{cases} {}^{CF}_{0}D^{\alpha}_{t}[f](t) + \sigma_0 f(t) + \sigma_1 f'(t) = g(t) \\ f(0) = 0 \end{cases}$$

and then apply the proposed method to recover f.

Choosing $-1 \leq j \leq 2$, it results in $\tilde{f} = \sum_{j=-1}^{2} \tilde{f}_j$. The plots of the f (blue) and $\tilde{f}$ (green) appear in figure.

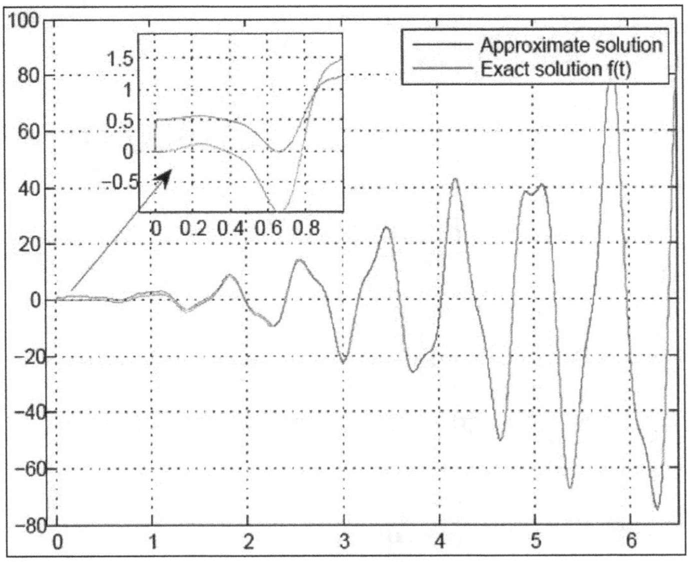

f and $\tilde{f}$.

This scheme can be adapted to solve boundary value problems. We show the procedure finding the solution to the fractional heat equation.

Boundary Value Problem

We show how to adapt the scheme used for initial value problem for solving boundary value problems with fractional partial differential equations in an example.

We will consider a fractional heat problem where we have replaced the classical time derivative by the Caputo-Fabrizio fractional derivative of order α:

$$\begin{pmatrix} {}^{CF}_{0}D^{\alpha}_{t}[u](x,t) - u_{xx}(x,t) = g(x,t), & x \in (0,1), t \in (0,T) \\ u(x,0) = 0 & x \in (0,1) \\ u(0,t) = u(1,t) = 0 & t \in (0,T) \end{pmatrix}$$

This equation models the evolution of temperatures in a bar of length 1, constituted by a heterogeneous material which has "memory," due to the fluctuations introduced by elements at different dimension scales.

The smooth and causal function g represents an external source. We look for smooth solutions $u \in C^2(0,1)\times(0,T)$ by separating variables and pose,

$$u(x,t) = \sum_{k\in\mathbb{Z}} u_k(t)\sin(k\pi x)$$

where $u_k(t)$ is the Fourier coefficients of $u(x,t)$ for each $t \in (0,T)$.

For the temporal part of the function, after replacing in equation $f(t) \approx \sum_{J_{min}}^{J_{max}} f_j(t) \approx \sum_{J_{min}}^{J_{max}} \sum_{k\in K_j} b_{jk}\psi_{jk}(t)$ we obtain an initial value problem like that described by equation $\begin{cases} {}^{CF}_{0}D^{\alpha}_{t}[f](t) + \sigma_0 f(t) + \sigma_1 f'(t) = g(t) \\ f(0) = 0 \end{cases}$ for each coefficient $u_k(t)$:

$$\begin{cases} D^{\alpha}_{0}[u_k](t) + (k\pi)^2 u_k(t) = B_k(t), t \in (0,T) \\ u_k(0) = 0 \end{cases}$$

with $B_k(t) = 2\int_0^1 g(x,t)\sin(k\pi x)dx$ and the Fourier coefficients of $g(x,t)$ for each $t \in (0,T)$. The uniqueness of solution is guaranteed because $u(x,0) = 0$, so $B_k(0)$ is null.

We show the approximate solution to equation,

$$\begin{pmatrix} {}^{CF}_{0}D^{\alpha}_{t}[u](x,t) - u_{xx}(x,t) = g(x,t), & x \in (0,1), t \in (0,T) \\ u(x,0) = 0 & x \in (0,1) \\ u(0,t) = u(1,t) = 0 & t \in (0,T) \end{pmatrix}$$

for $\alpha = 0.5, T = 3$, and $\begin{cases} {}^{CF}_{0}D^{\alpha}_{t}[f](t) + \sigma_0 f(t) + \sigma_1 f'(t) = g(t) \\ f(0) = 0 \end{cases}$ with v a smooth window.

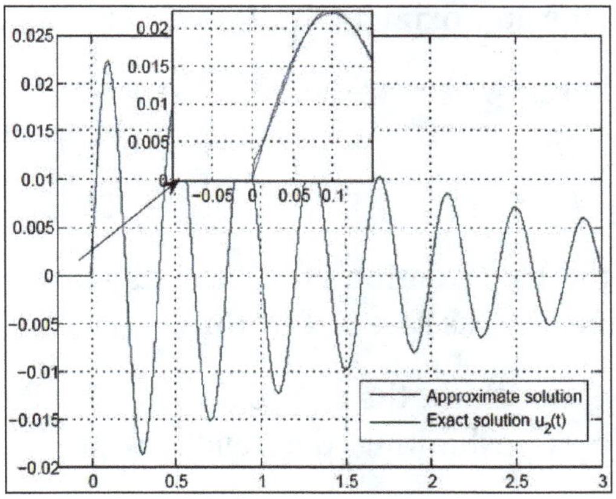

The approximate and the exact solutions to equation:

$$\begin{pmatrix} {}^{CF}_{0}D_t^{\alpha}[u](x,t)-u_{xx}(x,t)=g(x,t), & x\in(0,1),t\in(0,T) \\ u(x,0)=0 & x\in(0,1) \\ u(0,t)=u(1,t)=0 & t\in(0,T) \end{pmatrix}$$

with α = 0.8, T = 3, and B 3 = v t e – t=2 sin 5 πt .

In this case we only need to solve equation $\{D_0^{\alpha}[u_k](t)+(k\pi)^2 u_k(t)=u_k(0)=0$ for $k=2$, with $B_2=v(t)e^{-t/2}\sin(5\pi t)$. Wavelet analysis indicates that the 95% of the energy of B2 is concentrated in subspaces W_1, W_2, and W_3, and we obtained the following condition numbers for the band matrices of equation $\mathscr{M}^j\mathbf{b}^j=\mathbf{c}^j$: $\text{cond}_{\infty}(\mathscr{M}^0)=1.1153$, $\text{cond}_{\infty}(\mathscr{M}^1)=1.0663$, $\text{cond}_{\infty}(\mathscr{M}^2)=1.0132$, $\text{cond}_{\infty}(\mathscr{M}^3)=1.0098$, and $\text{cond}_{\infty}(\mathscr{M}^4)$ $=1.0076$. We consider levels $0\le j\le 4$ for the reconstruction, and the mean square error in this case is $\|u_2-\tilde{u}_2\|_{L_2}=3.502010^{-4}$.

$$\begin{cases} D_0^{\alpha}[u_k](t)+(k\pi)^2 u_k(t)=B_k(t),t\in(0,T) \\ u_k(0)=0 \end{cases}$$

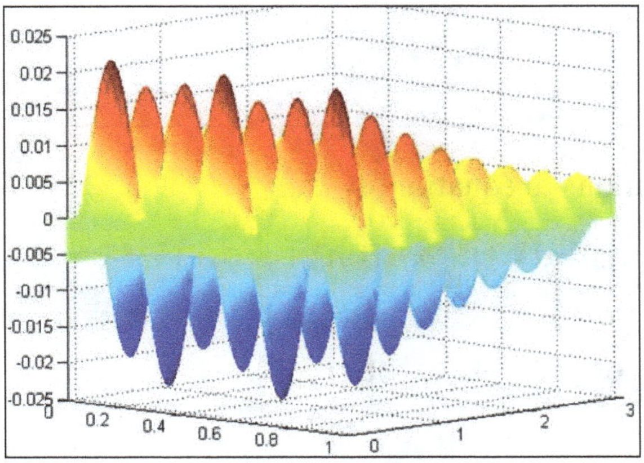

The approximate solutions to equation $\mathscr{M}^j\mathbf{b}^j=\mathbf{c}^j$ with α = 0.8, T = 3, g x t = v t e – t/2 sin 5 πt sin 2 πx, and v a smooth window in o 3.

In figure above we draw the approximate solution u of the heat problem described by equation $\begin{pmatrix} {}^{CF}_{0}D^{\alpha}_{t}[u](x,t)-u_{xx}(x,t)=g(x,t), & x\in(0,1),t\in(0,T) \\ u(x,0)=0 & x\in(0,1) \\ u(0,t)=u(1,t)=0 & t\in(0,T) \end{pmatrix}$ and in figure below the difference between the true solution u(x,t) and its approximation is plotted. The mean square error obtained in this case is 4.001610^{-4}.

Finally in figure below, we show the approximate solutions u^2 for different orders of derivation α, α→1, exhibiting the tendency to the solution of equation $\begin{cases} D^{\alpha}_{0}[u_k](t)+(k\pi)^2 u_k(t)=B_k(t), t\in(0,T) \\ u_k(0)=0 \end{cases}$ with $\alpha=1$, as expected (for $\alpha=1$, equation $\begin{pmatrix} {}^{CF}_{0}D^{\alpha}_{t}[u](x,t)-u_{xx}(x,t)=g(x,t), & x\in(0,1),t\in(0,T) \\ u(x,0)=0 & x\in(0,1) \\ u(0,t)=u(1,t)=0 & t\in(0,T) \end{pmatrix}$ describes classical heat problem, for a bar made of a homogeneous material).

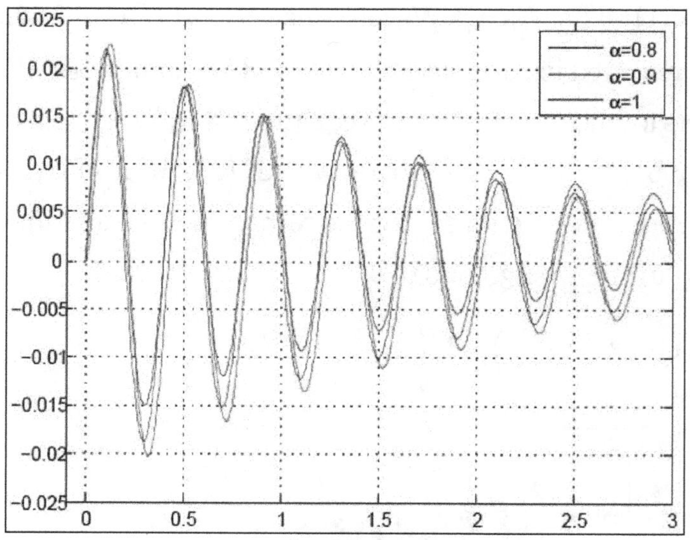

The approximate solutions to equation,

$$\begin{pmatrix} {}^{CF}_{0}D^{\alpha}_{t}[u](x,t)-u_{xx}(x,t)=g(x,t), & x\in(0,1),t\in(0,T) \\ u(x,0)=0 & x\in(0,1) \\ u(0,t)=u(1,t)=0 & t\in(0,T) \end{pmatrix}$$

with $\alpha \to 1$.

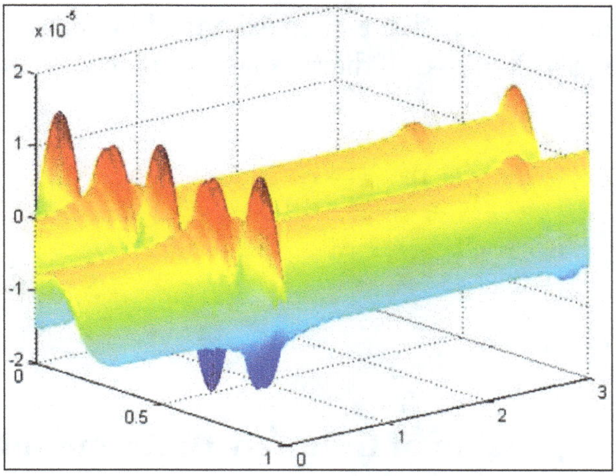

The difference between the true solution u x t to equation $\mathscr{M}^j \mathbf{b}^j = \mathbf{c}^j$ and its approximation by means of the wavelet scheme.

PERTURBED DIFFERENTIAL EQUATIONS WITH SINGULAR POINTS

Symbols $O, o, \sim$. Asymptotic Expansions of Functions

Let a function f(x) and $\phi(x)$ be defined in a neighborhood of $x = 0$.

Definition: If $\lim\limits_{x\to 0}\frac{f(x)}{\phi(x)} = M$, then write $f(x) = O(\phi(x)),\ x \to 0$, and M is constant.

If $\lim\limits_{x\to 0}\frac{f(x)}{\phi(x)} = 0$, then write $f(x) = o(\phi(x)),\ x \to 0$.

If $\lim\limits_{x\to 0}\frac{f(x)}{\phi(x)} = 0$, then write $f(x) \sim \phi(x),\ x \to 0$.

Definition: The sequence $\{\delta_n(\varepsilon)\}$, where $\delta_n(\varepsilon)$ defined in some neighborhood of zero, is called the asymptotic sequence in $\varepsilon \to 0$ if,

$$\lim_{\varepsilon\to 0}\frac{\delta_{n+1}(\varepsilon)}{\delta_n(\varepsilon)} = 0,\ \forall n = 1,2,\ldots$$

For example:

$$\{\varepsilon^n\},\{(1/\ln(1/\varepsilon))^n\},\{(\varepsilon\ln(1/\varepsilon))^n\}.$$

Everywhere below ε denotes a small parameter.

Definition: We say that $f(x)$ function can be expanded in an asymptotic series by the asymptotic sequence $\{\phi_n(x)\}, x \to 0$, if there exists a sequence of numbers $\{f_n\}$ and has the relation,

$$f(x) = \sum_{k=0}^{n} f_{k\phi k}(x) + O(\phi_{n+1}(x)), \quad x \to 0,$$

and write,

$$f(x) \sim \sum_{k=0}^{\infty} f_{k\phi k}(x), \; x \to 0,$$

The Asymptotic Expansion of Infinitely Differentiable Functions

Theorem Taylor and Maclaurin: If the function $f(x) \in C^{\infty}$ in some neighborhood of $x = 0$, then it can be expanded in an asymptotic series for the asymptotic sequence $\{x^n\}$, i.e.

$$f(x) \sim \sum_{n=1}^{\infty} f_n x^n, \text{ where } f_n = f^{(n)}(0)/n!.$$

Thus, the concept of an asymptotic expansion was given for the first time by Taylor and Maclaurin, although an explicit definition was given by Poincaré in 1886.

The Asymptotic Expansion of the Solution of the Ordinary Differential Equation

Consider the Cauchy problem for a normal ordinary differential equation,

$$y'(x) = f(x,y,\varepsilon), \quad y(0) = 0.$$

The function $f(x,y,\varepsilon)$ is infinitely differentiable on the variables x, y, ε in some neighborhood $O(0,0,0)$.

Theorem: The solution $y = y(x,\varepsilon)$ of equation $y'(x) = f(x,y,\varepsilon)$, $y(0) = 0$ exists and unique in some neighborhood point $O(0,0,0)$ and $y(x,\varepsilon) \in C^{\infty}$, for small x,ε.

Corollary: The solution of equation $y'(x) = f(x,y,\varepsilon)$, $y(0) = 0$ can be expanded in an asymptotic series by the small parameter ε, i.e.

$$y(x,\varepsilon) = \sum_{k=1}^{\infty} \varepsilon^k y_k(x).$$

Here and below, the equality is understood in an asymptotic sense.

Theorem above for the case when $f(x,y,\varepsilon)$ is analytical was given in equation $y'(x) = f(x,y,\varepsilon)$, $y(0) = 0$ by Duboshin. This theorem is not true if $f(x,y,\varepsilon)$ is not smooth at ε. For example, the solution of a singularly perturbed equation,

$$\varepsilon y'(x) = -y(x), \quad y(0) = a,$$

function $y(x) = ae^{-x/\varepsilon}$ and is not expanded in an asymptotic series in powers of ε, because here $f(x,y,\varepsilon) = -y(x)/\varepsilon$ and f have a pole of the first order with respect to ε.

For example: Series,

$$y(x,\varepsilon) = 1 + \varepsilon x^{-1} + (\varepsilon x^{-1})^2 + ... + (\varepsilon x^{-1})^n + ...$$

It is not uniform valid asymptotic series on the interval [0, 1], but it is a uniform valid asymptotic expansion of the segment $[\varepsilon^{\alpha},1]$, where $0 < \alpha < 1$.

Singularly Perturbed Ordinary Differential Equations

We divide such equations into three types:

Singular perturbations of ordinary differential equations such as the Prandtl-Tikhonov i.e., perturbed equations that contain a small parameter at the highest derivative, i.e., equations of the form,

$$y'(x) = f(x,y,\varepsilon), \quad y(0) = 0, \quad \varepsilon z'(x) = g(x,y,\varepsilon), \quad z(0) = 0,$$

where f, g are infinitely differentiable in the variables x,y,ε in the neighborhood of O(0, 0, 0). It is obvious that unperturbed equation (ε = 0),

$$y_0'(x) = f(x,y,0), \quad 0 = g(x,y,0)$$

is a first order.

Definition: Singularly perturbed equation will be called bisingulary perturbed if the corresponding unperturbed differential equation has a singular point, or this one is an unbounded solution in the considering domain.

For example:

Equation $\varepsilon y'(x) = -y(x)$ is a singularly perturbed ordinary differential equation.

Equation Vander Pol,

$$\varepsilon y''(x) + (1 - y^2(x))y'(x) + y(x) = 0.$$

It is a bisingularly perturbed ordinary differential equation with singular points, if:

$$y(x) = \pm 1.$$

$\varepsilon y'(x) - xy(x) = 1, x \in [0,1]$ is a bisingularly perturbed equation, because the unperturbed equation has an unbounded solution $y_0(x) = -x^{-1}$.

$\varepsilon y''(x) - xy(x) = 1, \quad x \in [0,1]$ is a bisingularly perturbed equation also.

Singularly perturbed differential equations such as the Lighthill's type, in which the order of the corresponding unperturbed equation is not reduced, but has a singular point in the considering domain.

For example, a Lighthill model equation,

$$(x + \varepsilon y(x))y'(x) + p(x)y(x) = r(x), \quad y(1) = a$$

where $x \in [0,1], \; p(x), r(x) \in C^{\infty}[0,1]$. For unperturbed equation,

$$xy'_0(x) + p(x)y_0(x) = r(x),$$

point $x = 0$ is a regular singular point.

A singularly perturbed equation with a small parameter is considered on an infinite interval. For example, the Lagerstrom equation:

$$y''(x) + nx^{-1}y'(x) + y(x)y'(x) = \beta(y'(x))^2,$$
$$y(\varepsilon) = 0, y(\infty) = 1.$$

Where $0 < \beta$ is a given number and n is the dimension space.

The division into such classes is conditional, because singularly perturbed equation of Van der Pol in the neighborhood of points y=±1 leads to an equation of Lighthill type.

Methods of Construction of Asymptotic Expansions of Solutions of Singularly Perturbed Differential Equations

The method of matching of outer and inner expansions is the most common method for constructing asymptotic expansions of solutions of singularly perturbed differential equations. Justification for this method is given by Il'in. However, this method is relatively complex for applied scientists.

The boundary layer function method (or composite asymptotic expansion) dates back to the work of many mathematicians. For the first time, this method for a singularly perturbed differential equations in partial derivatives is developed by Vishik and Lyusternik and for nonlinear integral-differential equations (thus for the ordinary differential equations) Imanaliev, O'Malley and Hoppenstedt.

It should be noted that, for the first time, the uniform valid asymptotic expansion of the solution of equation $y(x) = \mu^{-1}\pi_{-1}(t) + Y_0(x) + \pi_0(t) + \mu(Y_1(x) + \pi_1(t)) + \mu^2(Y_2(x) + \pi_2(t)) + \ldots, \mu \to 0$, is constructed by Vasil'eva after Wasow and Sibuya in 1963 by the method of matching.

- This method is constructive and understandable for the applied scientists.

- The method of Lomov or regularization method is applied for the construction of uniformly valid solutions of a singularly perturbed equation and will apply Fredholm ideas.
- The method WKB or Liouville-Green method is used for the second-order differential equations.
- The method of multiple scales.
- The averaging method is applicable to the construction of solutions of a singularly perturbed equation on a large but finite interval.

Here, we consider a bisingularly perturbed differential equations and types of equations of Lighthill and Lagerstrom. Here, we generalize the boundary layer function method for bisingular perturbed equations. We will construct a uniform asymptotic solution of the Lighthill model equation by the method of uniformization and construct the explicit solution of this one by the generalized method of the boundary layer functions.

Furthermore, we construct a uniformly convergent solution of the Lagerstrom model equation by the method of fictitious parameter.

Bisingularly Perturbed Ordinary Differential Equations

Singularly Perturbed of the First-order Equation with a Turning Point

Consider the Cauchy problem:

$$\varepsilon y'(x) + xy(x) = f(x), \quad 0 < x \le 1, \quad y(0) = a,$$

where $f(x) \in C^{\infty}[0,1]$, $f(x) = \sum_{k=0}^{\infty} f_k x^k$, $f_k = f^{(k)}(0)/k!, f_0 \neq 0$; a is the constant.

Explicit solution of the equation above has the form:

$$y(x) = ae^{-x^2/2\varepsilon} + \frac{1}{\varepsilon}\int_0^x e^{(s^2-x^2)/2\varepsilon} f(s)ds.$$

The corresponding unperturbed equation ($\varepsilon = 0$),

$$-x\tilde{y}(x) + f(x) = 0,$$

has a solution $\tilde{y}(x) = f(x)/x$, which is unbounded at $x = 0$.

If you seek a solution to equation $y'(x) = f(x,y,\varepsilon)$, $y(0) = 0$ in the form,

$$y(x) = y_0(x) + \varepsilon y_1(x) + \varepsilon^2 y_2(x) + \ldots,$$

then,

$$y_0(x) = \frac{f(x)}{x} \sim f_0 x^{-1}, \quad x \to 0,$$

$$y_1(x) = x^{-1} y'_0(x) \sim f_0 x^{-3}, \quad x \to 0,$$

$$y_2(x) = x^{-1} y'_1(x) \sim 3 f_0 x^{-5}, \quad x \to 0,$$

$$y_3(x) = x^{-1} y'_2(x) \sim 3 \cdot 5 f_0 x^{-5}, \quad x \to 0,$$

$$y_n(x) = x^{-1} y'_{n-1}(x) \sim 3 \cdot 5 \cdot \ldots \cdot (2n-1) f_0 x^{-(2n+1)}, \quad x \to 0,$$

and a series of equation $y(x) = y_0(x) + \varepsilon y_1(x) + \varepsilon^2 y_2(x) + \ldots$, is asymptotic in the segment $(\sqrt{\varepsilon}, 1]$, and the point $x_0 = \sqrt{\varepsilon} = \mu$ is singular point of the asymptotic series of equation $y(x) = y_0(x) + \varepsilon y_1(x) + \varepsilon^2 y_2(x) + \ldots,$. Therefore, the solution of equation $\varepsilon y'(x) + xy(x) = f(x),\ 0 < x \le 1,\ y(0) = a$, we will seek in the form,

$$y(x) = \mu^{-1}\pi_{-1}(t) + Y_0(x) + \pi_0(t) + \mu\left(Y_1(x) + \pi_1(t)\right) + \mu^2\left(Y_2(x) + \pi_2(t)\right) + \ldots, \ \mu \to 0,$$

where $Y_k(x) \in C^{(\infty)}[0,1]$, $\pi_k(t) \in C^{(\infty)}[0, \mu^{-1}]$, $x = \mu t$ and boundary layer functions $\pi_k(t)$ decreasing by power law as $t \to \infty$, that is, $\pi_k(t) = O(t^{-m}),\ t \to \infty, m \in N$.

Substituting equation,

$$y(x) = \mu^{-1}\pi_{-1}(t) + Y_0(x) + \pi_0(t) + \mu\left(Y_1(x) + \pi_1(t)\right) + \mu^2\left(Y_2(x) + \pi_2(t)\right) + \ldots, \ \mu \to 0,$$

into equation $\varepsilon y'(x) + xy(x) = f(x),\ 0 < x \le 1,\ y(0) = a$, we obtain,

$$\begin{aligned}&\pi'_{-1}(t) + \mu^2 Y'_0(x) + \mu\pi'_0(t) + \mu^3 Y'_1(x) + \mu^2\pi'_1(t) + \mu^4 Y'_2(x) + \mu^3\pi'_2(t) + \mu^5 Y'_3(x) + \mu^4\pi'_3(t) + \ldots \\ &+ xY_0(x) + \mu x Y_1(x) + \mu^2 x Y_2(x) + \mu^3 x Y_3(x) + \ldots + t\pi_{-1}(t) + \mu t\pi_0(t) + \mu^2 t\pi_1(t) + \mu^3 t\pi_2(t) \\ &+ \mu^4 t\pi_3(t) + \ldots = f(x).\end{aligned}$$

The initial conditions for the functions $\pi_{k-1}(t),\ k = 0,1,\ldots$ we take in the next form,

$$\pi_{-1}(0) = 0, \pi_0(0) = a - Y_0(0), \pi_k(0) = -Y_k(0), \quad k = 1,2,\ldots$$

From equation,

$$\begin{aligned}&\pi'_{-1}(t) + \mu^2 Y'_0(x) + \mu\pi'_0(t) + \mu^3 Y'_1(x) + \mu^2\pi'_1(t) + \mu^4 Y'_2(x) + \mu^3\pi'_2(t) + \mu^5 Y'_3(x) + \mu^4\pi'_3(t) + \ldots \\ &+ xY_0(x) + \mu x Y_1(x) + \mu^2 x Y_2(x) + \mu^3 x Y_3(x) + \ldots + t\pi_{-1}(t) + \mu t\pi_0(t) + \mu^2 t\pi_1(t) + \mu^3 t\pi_2(t) \\ &+ \mu^4 t\pi_3(t) + \ldots = f(x).\end{aligned}$$

we have,

$$\mu^0 : \pi'_{-1}(t) + t\pi_{-1}(t) + xY_0(x) = f(x),$$

$$\mu^1 : \pi'_0(t) + t\pi_0(t) + xY_1(x) = 0,$$

$$\mu^{k+1} : \quad \pi'_k(t) + t\pi_k(t) + xY_{k+1}(x) + Y'_{k-1}(x) = 0, \quad k = 1, 2, \ldots$$

To $Y_0(x)$ function has been smooth, and we define it from the equation,

$$xY_0(x) = f(x) - f_0 \Rightarrow Y_0(x) = (f(x) - f_0)/x,$$

and then from equation $\mu^0 : \pi'_{-1}(t) + t\pi_{-1}(t) + xY_0(x) = f(x),$ we have obtained the equation,

$$\pi'_{-1}(t) + t\pi_{-1}(t) = f_0$$

Therefore,

$$\pi_{-1}(t) = f_0 e^{-t^2/2} \int_0^t e^{s^2/2} ds \in C^\infty[0, \mu^{-1}],$$ y

Obviously, this function bounded and is infinitely differentiable on the segment $[0, \mu^{-1}]$, and,

$$\pi_{-1}(t) = -\frac{f_0}{t}\left(1 + \frac{1}{t^2} + \frac{3}{t^4} + \ldots\right), \quad t \to \infty.$$

This asymptotic expression can be obtained by integration by parts the integral expression for $\pi_{-1}(t)$.

Equation $\mu^1 : \pi'_0(t) + t\pi_0(t) + xY_1(x) = 0,$ define $Y_1(x)$ and $\pi_0(t)$. Let $Y_1(x) \equiv 0$, then

$$\pi'_0(t) + t\pi_0(t) = 0, \; \pi_0(0) = a - f_1.'$$

Hence, we find,

$$\pi_0(t) = (a - f_1)e^{-t^2/2}.$$

From equation $\mu^{k+1} : \quad \pi'_k(t) + t\pi_k(t) + xY_{k+1}(x) + Y'_{k-1}(x) = 0, \quad k = 1, 2, \ldots$ for k=1, we have,

$$\pi'_1(t) + t\pi_1(t) + xY_2(x) + Y'_0(x) = 0.$$

Let $xY_2(x) = Y'_0(0) - Y'_0(x)$, then $\pi'_1(t) + t\pi_1(t) = -Y'_0(0)$.

From these, we get,

$$Y_2(x)=(Y'_o(o)-Y'_o(x))/x, \quad \pi_1(t)=-f_2e^{-t^2/2}\int_o^t e^{s^2/2}ds \in C^\infty[o,\mu^{-1}],$$

and

$$\pi_{-1}(t)=\frac{f_2}{t}\left(1+\frac{1}{t^2}+\frac{3}{t^4}+\dots\right), \quad t\to\infty.$$

From equation $\mu^{k+1}: \ \pi'_k(t)+t\pi_k(t)+xY_{k+1}(x)+Y'_{k-1}(x)=o, \quad k=1,2,\dots$ for k = 2, we have,

$$\pi'_2(t)+t\pi_2(t)+xY_3(x)+Y'_1(x)=o \text{ or } \pi'_2(t)+t\pi_2(t)+xY_3(x)=o.$$

Let $Y_3(x)\equiv o$, then,

$$\pi'_2(t)+t\pi_2(t)=o, \ \pi_2(o)=-Y_2(o)=2f_3.$$

From this, we get,

$$\pi_2(t)=2f_3e^{-t^2/2}.$$

Analogously continuing this process, we determine the others of the functions:

$$Y_K(x), \pi_k(t).$$

In order to show that the constructed series of equation $y(x)=\mu^{-1}\pi_{-1}(t)+Y_o(x)+\pi_o(t)+\mu\left(Y_1(x)+\pi_1(t)\right)+\mu^2\left(Y_2(x)+\pi_2(t)\right)+\dots, \ \mu\to o,$ is asymptotic series, we consider remainder term $R_m(x)=y(x)-y_m(x),$

where,

$$y_m(x)=\frac{1}{\mu}\pi_{-1}(t)+Y_o(x)+\pi_o(t)+\mu(Y_1(x)+\pi_1(t))+\dots+\mu^m(Y_m(x)+\pi_m(t)).$$

For the remainder term $R_m(x)$, we obtain a problem:

$$\mathring{a}R'_m(x)+xR_m(x)=-\mu^{m+2}\,Y'_m(x), \ o<x\le 1, \ R_m(o)=o.$$

We note that if m is odd, then $Y'_m(x)\equiv o.$

$$R_m(x)=-\mu^m e^{-x^2/2\varepsilon}\int_o^x Y'm(s)e^{s^2/2\varepsilon}ds,$$

and from this, we have,

$$R_m(x) = O(\mu^m), \mu \to 0, x \in [0,1].$$

Bisingularly Perturbed in a Homogenous Differential Equation of the Airy Type

Consider the boundary value problem for the second-order ordinary in a homogenous differential equation with a turning point,

$$\varepsilon y''(x) - xy(x) = f(x), \quad x \in (0,1),$$

$$y(0) = 0, \quad y(1) = 0.$$

where,

$$f(x) = \sum_{k=0}^{\infty} f_k x^k, \; x \to 0, \; f_k = f^{(k)}(0)/k!, \; f_0 \neq 0.$$

Without loss of generality, we consider the homogeneous boundary conditions, since $y(0) = a, y(1) = b, \; a^2 + b^2 \neq 0$, using transformation,

$$y(x) = a + (b-a)x + z(x),$$

can lead to conditions.

If the asymptotic solution of the equation $\varepsilon y''(x) - xy(x) = f(x), \; x \in (0,1)$, to $y(0) = 0, \; y(1) = 0$ we seek in the form

$$y(x) = y_0(x) + \varepsilon y_1(x) + \varepsilon^2 y_2(x) + \ldots,$$

then we have,

$$y_0(x) = -\frac{f(x)}{x} \sim f_0 x^{-1}, \; x \to 0,$$

$$y_1(x) = x^{-1} y''_0(x) \sim 1 \cdot 2 f_0 x^{-4}, \quad x \to 0,$$

$$y_2(x) = x^{-1} y''_1(x) \sim 1 \cdot 2 \cdot 4 \cdot 5 \, f_0 x^{-7}, \quad x \to 0,$$

$$y_3(x) = x^{-1} y''_2(x) \sim 1 \cdot 2 \cdot 4 \cdot 5 \cdot 7 \cdot 8 \, f_0 x^{-10}, \quad x \to 0,$$

$$y_n(x) = x^{-1} y''_{n-1}(x) \sim 1 \cdot 2 \cdot 4 \cdot 5 \cdot 7 \cdot 8 \cdot \ldots \cdot (3n-2) \cdot (3n-1) f_0 x^{-(3n+1)}, 0 < n, \quad x \to 0,$$

and the series $y(x) = y_0(x) + \varepsilon y_1(x) + \varepsilon^2 y_2(x) + \ldots$, is asymptotic in the segment $(\sqrt[3]{\varepsilon}, 1]$.

The point $x_0 = \sqrt[3]{\varepsilon} = \mu$ is singular point of asymptotic series.

The solution of equations $\varepsilon y''(x)-xy(x)=f(x),\ x\in(0,1)$, and $y(0)=0,\ y(1)=0$ will be sought in the form,

$$y(x)=\mu^{-1}\pi_{-1}(t)+\sum_{k=0}^{\infty}\mu^{k}\left(Y_k(x)+\pi_k(t)\right)+\sum_{k=0}^{\infty}\lambda^{k}w_k(\eta),$$

where $t=x/\mu,\ \mu=\sqrt[3]{\varepsilon},\ \varsigma=(1-x)/\lambda,\lambda=\sqrt{\varepsilon}$. Here, $Y_k(x)\in C^{\infty}[0,1],\ \pi_k(t)\in C^{\infty}[0,1/\mu]$ is boundary layer function in a neighborhood of $t=0$ and decreases by the power law as $t\to\infty$, and the function $w_k(t)\in C^{\infty}[0,1/\lambda]$ is boundary function in a neighborhood of $\eta=0$ and decreases exponentially as $\eta\to\infty$.

Substituting equation $y(x)=\mu^{-1}\pi_{-1}(t)+\sum_{k=0}^{\infty}\mu^{k}\left(Y_k(x)+\pi_k(t)\right)+\sum_{k=0}^{\infty}\lambda^{k}w_k(\eta)$, in equation $\varepsilon y''(x)-xy(x)=f(x),\ x\in(0,1)$, we get,

$$\sum_{k=0}^{\infty}\mu^{k}(\pi''_{k-1}(t)-t\pi_{k-1}(t))+\sum_{k=0}^{\infty}\mu^{k+3}Y''_k(x)-x\sum_{k=0}^{\infty}\mu^{k}\,Y_k(x)=f(x)$$

$$\sum_{k=0}^{\infty}\lambda^{k}\left(w''_k(\eta)-(1-\lambda\eta)w_k(\eta)\right)=0.$$

From equation $\sum_{k=0}^{\infty}\mu^{k}(\pi''_{k-1}(t)-t\pi_{k-1}(t))+\sum_{k=0}^{\infty}\mu^{k+3}Y''_k(x)-x\sum_{k=0}^{\infty}\mu^{k}\,Y_k(x)=f(x)$ we have,

$$\mu^{0}:\pi''_{-1}(t)-t\pi_{-1}(t)-xY_0(x)=f(x),$$

$$\mu^{1}:\pi''_{0}(t)-t\pi_{0}(t)-xY_1(x)=0,$$

$$\mu^{2}:\pi''_{1}(t)-t\pi_{1}(t)-xY_2(x)=0,$$

$$\mu^{3}:\pi''_{2}(t)-t\pi_{2}(t)+Y''_{0}(x)-xY_3(x)=0,$$

$$\mu^{k}:\pi''_{k-1}(t)-t\mu_{k-1}(t)+Y''_{k-3}(x)-xY_k(x)=0,\ k>3,$$

Boundary conditions for functions $\pi_{k-1}(t), k=0,1,\ldots$ we take next form,

$$\pi_{-1}(0)=0,\pi_k(0)=-Y_k(0),\lim_{\mu\to 0}\pi_{k-1}(1/\mu)=0,\ k=0,1,2,\ldots$$

To $Y_0(x)$ function has been smooth; therefore, we define it from the equation,

$$-xY_0(x)=f(x)-f_0\Rightarrow Y_0(x)=-(f(x)-f_0)/x,$$

then from equation $\mu^{0}: \pi''_{-1}(t)-t\pi_{-1}(t)-xY_{0}(x)=f(x)$, we have the equation,

$$\pi''-1^{(t)-t}\pi_{-1}(t)=f_{0}^{\cdot}$$

Let us prove an auxiliary lemma.

Lemma: Next boundary value problem,

$$z''(t)-tz\ t\ =b,\quad 0<t<1/\mu,\ \text{here b is the constan}$$

$$z(0)=z^{0},\ z(1/\mu)\to 0,\ \mu\to 0$$

will have the unique solution and this one have next form,

$$z(t)=z^{0}\frac{Ai(t)}{Ai(0)}-\pi b\left(Ai(t)\int_{0}^{t}Bi(s)ds+Bi(t)\int_{t}^{1/\mu}Ai(s)ds-Ai(t)\sqrt{3}\int_{0}^{1/\mu}Ai(s)ds\right),$$

and $z(t)\in C^{\infty}[0,\mu^{-1}]$.

Proof: We verify the boundary conditions:

$$z(0)=z^{0}-\pi b\left(Bi(0)\int_{t}^{1/\mu}Ai(s)ds-Ai(0)\sqrt{3}\int_{0}^{1/\mu}Ai(s)ds\right),$$

as $Bi(0)=Ai(0)\sqrt{3}$, so $z(0)=z^{0}$.

$$z(1/\mu)=z^{0}\frac{Ai(1/\mu)}{Ai(0)}-\pi b\left(1-\sqrt{3}\right)Ai(1/\mu)\int_{t}^{1/\mu}Bi(s)ds,$$

as $Ai(t)\sim t^{-1/4}e^{-\frac{2}{3}t^{3/2}}$, $Bi(t)\sim t^{-1/4}e^{\frac{2}{3}t^{3/2}}$, $t\to\infty$, so $z(1/\mu)=O(\mu)$, $\mu\to 0$.

Now we show that z(t) satisfies equation $z''(t)-tz(t)=b,\ 0<t<1/\mu,\ \text{here b is the constant}$. For this, we compute derivatives:

$$z'(t)=z^{0}\frac{Ai'(t)}{Ai(0)}-\pi b\left(Ai'(t)\int_{0}^{t}Bi(s)ds+Bi'(t)\int_{t}^{1/\mu}Ai(s)-Ai'(t)\sqrt{3}\int_{0}^{1/\mu}Ai(s)ds\right)$$

$$z''(t)=z^{0}\frac{Ai''(t)}{Ai(0)}-\pi b\left(Ai''(t)\int_{0}^{t}Bi(s)ds+Bi''(t)\int_{t}^{1/\mu}Ai(s)ds-\frac{1}{\pi}-Ai''(t)\sqrt{3}\int_{0}^{1/\mu}Ai(s)ds\right)$$

Substituting the expressions for z''(t) and z(t) in equation $z(0)=z^{0},\ z(1/\mu)\to 0,\ \mu\to 0$ and given that $Ai''(t)-t\,Ai(t)\equiv 0$ and $Bi''(t)-t\,Bi(t)\equiv 0$, we get: $b\equiv b$.

The uniqueness of z(t) the solution is proved by contradiction. Let u(t) also be a solution of equation $z''(t)-tz(t)=b,\ 0<t<1/\mu,\ \text{here b is the constant}$ and

$z(0)=z^0$, $z(1/\mu)\to 0$, $\mu\to 0$, $z(t)\neq u(t)$. Considering the function $r(t)=z(t)-u(t)$, for the function r(t), we obtain the problem,

$$r''(t)-tr(t)=0,\quad 0<t<1/\mu,\quad r(0)=0,\quad r(1/\mu)\to 0,\quad \mu\to 0.$$

The general solution of the homogeneous equation is,

$$r(t)=c_1\,Ai(t)+c_2\,Bi(t);\quad c_{1,2}\text{ is the constant.}$$

Considering the boundary condition $r(1/\mu)\to 0,\mu\to 0$, we have $c_2=0$; $r(t)=c_1\,Ai(t)$. And the second condition $r(0)=0, c_1=0$ follows. This implies that $r(t)\equiv 0$.

Therefore, $z(t)\equiv u(t)$. It is obvious that $z(t)\in C^\infty[0,\mu^{-1}]$. Auxiliary lemma is proved.

This Lemma above implies the existence and uniqueness of $\pi_{-1}(t)\in C^\infty[0,\mu^{-1}]$ solution of the problem:

$$\pi''_{-1}(t)-t\pi_{-1}(t)=f_0,\quad 0<t<1/\mu,\quad \pi_{-1}(0)=0,\quad \pi_{-1}(1/\mu)\to 0,\quad \mu\to 0.$$

This function bounded and is infinitely differentiable on the segment $[0,\mu^{-1}]$, and as $t\to\infty$:

$$\pi_{-1}(t)=-\frac{f_0}{t}\left(1+\frac{1\cdot 2}{t^3}+\frac{1\cdot 2\cdot 4\cdot 5}{t^6}+\ldots\right).$$

This asymptotic expression can be obtained by integration by parts the integral expression for $\pi_{-1}(t)$.

From equation $\mu^1:\pi''_0(t)-t\pi_0(t)-xY_1(x)=0$, we define $Y_1(x)$ and $\pi_0(t)$. Let $Y_1(x)\equiv 0$, then,

$$\pi''_0(t)-t\pi_0(t)=0,\quad \pi_0(0)=f_1,\quad \pi_0(1/\mu)\to 0,\quad \mu\to 0,$$

And by above given Lemma, we have,

$$\pi_0(t)=f_1 Ai(t)/Ai(0).$$

Analogously, from equation $\mu^2:\pi''_1(t)-t\pi_1(t)-xY_2(x)=0$, we define $Y_2(x)$ and $\pi_1(t)$. Let $Y_2(x)\equiv 0$, then,

$$\pi''_1(t)-t\pi_1(t)=0,\quad \pi_1(0)=0,\quad \pi_0(1/\mu)\to 0,\quad \mu\to 0.$$

In view of above given Lemma, we have $\pi_1(t)\equiv 0$.

To $Y_3(x)$ function has been smooth; as above, we define it from the equation,

$$xY_3(x)=Y''_0(x)-Y''_0(0)\Rightarrow Y_3(x)=(Y''_0(x)-Y''_0(0))/x,\quad (Y''_0(0)=-2f_3),$$

then equation $\mu^3 : \pi''_2(t) - t\pi_2(t) + Y''_0(x) - xY_3(x) = 0$, to $\pi_2(t)$ hase the problem,

$$\pi''_2(t) - t\pi_2(t) = 2f_3 \quad \pi_2(0) = 0, \quad \pi_2(1/\mu) \to 0, \quad \mu \to 0.$$

By above given Lemma, we can write an explicit solution to this problem, and this solution bounded and is infinitely differentiable on the segment $[0, \mu^{-1}]$, and as $t \to \infty$:

$$\pi_2(t) = -\frac{2f_3}{t}\left(1 + \frac{1 \cdot 2}{t^3} + \frac{1 \cdot 2 \cdot 4 \cdot 5}{t^6} + \ldots\right).$$

Analogously continuing this process, we determine the rest of the functions $Y_k(x), \pi_k(t)$.

Now we will define functions $w_k(\eta)$ from the equality $\sum_{k\ 0}^{\infty} \lambda \ (w''\ (\eta) - (1 - \lambda\eta) w\ (\eta)) = 0$ by using the boundary conditions $y(1) = 0$. We state problems,

$$Lw_0 \equiv w''_0(\eta) - w_0(\eta) = 0, \ w_0(0) = Y_0(1), \ \lim_{\eta \to \infty} w_0(\eta) = 0$$

$$Lw_k \equiv \eta w_{k-1}(\eta), w_{2i}(0) = Y_{3i}(1), \ w_{2i-1}(0), \ \lim_{\eta \to \infty} w_k(\eta) = 0, \quad k, i \in N.$$

One can easily make sure that all these equations $Lw_0 \equiv w''_0(\eta) - w_0(\eta) = 0$, $w_0(0) = Y_0(1)$, $\lim_{\eta \to \infty} w_0(\eta) = 0$and $Lw_k \equiv \eta w_{k-1}(\eta), w_{2i}(0) = Y_{3i}(1)$, $w_{2i-1}(0)$, $\lim_{\eta \to \infty} w_k(\eta)$ $= 0$, $k, i \in N$. have unique solutions such that $w_k(\eta) \in C^\infty[0, \infty)$, $w_k(\eta) = O(e^{-\eta})$, with $\eta \to \infty$.

Thus, all functions $Y_k(x), w_k(\eta)$, and $\pi_k(t)$ in equality $y(x) = \mu^{-1}\pi_{-1}(t) \ldots \sum_{k=0}^{\infty} \lambda^k w_k(\eta)$, are defined, i.e., a formally asymptotic expansion is constructed. Let us justify the constructed expansion. Let,

$$y_m(x) = \mu^{-1}\pi_{-1}(t) + \sum_{k=0}^{3m} \mu^k (Y_k(x) + \pi_k(t)) + \sum_{k=0}^{2m} \lambda^k w_k(\eta), \quad r_m(x) = y(x) - y_m(x).$$

Then for the remainder term, we state the following problem:

$$\varepsilon r''_m(x) - x r_m(x) = O(\varepsilon^{m+1/2}), \quad \varepsilon \to 0, \ x \in (0,1).$$

$$r_m(0) = O(e^{-1\sqrt{\varepsilon}}), \quad r_m(1) = O(\varepsilon^{m+1}) \quad \varepsilon \to 0.$$

Let $r_m(x) = (2 - x^2) R_m(x)/2$, and then equations above take the form,

$$\varepsilon R''_m(x) - \frac{4x\varepsilon}{2 - x^2} R'_m(x) - \left(\frac{2\varepsilon}{2 - x^2} + x\right) R_m(x) = O(\varepsilon^{m+1/2}), \quad \varepsilon \to 0,$$

$$R_m(0) = O(e^{-1\sqrt{\varepsilon}}), \ R_m(1) = O(\varepsilon^{m+1}), \quad \varepsilon \to 0.$$

According to the maximum principle, we have $R_m(x) = O(\varepsilon^{m-1/2}),\ \varepsilon \to 0, x \in [0,1]$.

Hence, we get $r_m(x) = O(\varepsilon^{m-1/2}),\ \varepsilon \to 0,\ x \in [0,1]$.

Thus, we have proved.

Theorem: Let $f(0) \neq 0$, then the solution to equation $\varepsilon y''(x) - xy(x) = f(x),\ x \in (0,1)$, and $y(0) = 0,\ y(1) = 0$ will have next form,

$$y(x) = \frac{1}{\sqrt[3]{\varepsilon}}\pi_{-1}\left(\frac{x}{\sqrt[3]{\varepsilon}}\right) + \sum_{k=0}^{\infty} \sqrt[3]{\varepsilon^k}\left(y_k(x) + \pi_k\left(\frac{x}{\sqrt[3]{\varepsilon}}\right)\right) + \sum_{k=0}^{\infty} \sqrt{\varepsilon^k} w_k\left(\frac{1-x}{\sqrt{\varepsilon}}\right).$$

Example: Consider the problem,

$$\varepsilon y''(x) - xy(x) = 1 + x,\quad x \in (0,1),\quad y(0) = 0,\quad y(1) = 0.$$

The asymptotic solution this problem we can represent in the form,

$$y(x) = \mu^{-1}\pi_{-1}(t) + \sum_{k=0}^{3}\mu^k\left(Y_k(x) + \pi_k(t)\right) + \omega_0(\eta) + \lambda w_1(\eta) + \lambda^2 w_2(\eta) + R(x)$$

We have got $Y_0(x) = -(1 + x - 1)/x = -1,\ Y_{1,2,3}(x) \equiv 0$,

$$\pi_{-1}(t) = -\pi\left(Ai(t)\int_0^t Bi(s)ds + Bi(t)\int_t^{1/\mu} Ai(s)ds - Ai(t)\sqrt{3}\int_0^{1/\mu} Ai(s)ds\right),$$

$$\pi_0(t) = Ai(t)/Ai(0),\ \pi_{1,2,3}(t) \equiv 0, w_0(\eta) = 2e^{-\eta},\ w_k(\eta) = O(e^{-\eta}),\ k = 1,2.$$

$$\varepsilon R''(x) - xR(x) = O(\varepsilon^{3/2}), 0 < x < 1, R(0) = O(e^{-1/\sqrt{\varepsilon}}), R(1) = O(\varepsilon^2),\ \varepsilon \to 0.$$

We have,

$$y(x) = \varepsilon^{-1/3}\pi_{-1}(t) - 1 + 2e^{-(1-x)/\sqrt{\varepsilon}} + \pi_0(t) + \sqrt{\varepsilon}w_1(\eta) + \varepsilon w_2(\eta) + O(\sqrt{\varepsilon}), \varepsilon \to 0.$$

Bisingularly Perturbed Equation of the Second Order with a Regularly Singular Point

Consider the boundary value problem,

$$L_\varepsilon y \equiv \varepsilon y'' + xy' - q(x)y = f(x),\ x \in [0,1],$$

$$y(0) = 0,\ y(1) = 0,$$

where $q(x), f(x) \in C^\infty[0,1]$.

Here, for simplicity, we consider the case $q(0)=1, q(x)\geq 1$.

The solution of the unperturbed problem,

$$My \equiv xy' - q(x)y = f(x),$$

represented as,

$$y_0(x) \equiv xp(x)\int_1^x r(s)s^{-2}ds,$$

where,

$$r(x) = p^{-1}(x)f(x), p(x) = \exp\left\{\int_1^x \left(q(x)^{-1}\right)s^{-1}ds\right\}.$$

Extracting in equation $y_0(x) \equiv xp(x)\int_1^x r(s)s^{-2}ds$, the main part of the integral in the sense of Hadamard, it can be represented as,

$$y_0(x) = a(x) + r_1 xp(x)\ln x,$$

where,

$$a(x) = xp(x)\int_1^x \left(r(s) - r_0 - r_1 s\right)s^{-2}ds + r_0 p(x)[x-1],$$

$$r_0 = r(0), r_1 = r'(0) = p(0)^{-1}\left[f'(0) - q'(0)f(0)\right].$$

Function $a(x) \in C^\infty[0,1]$.

Theorem: Suppose that the conditions referred to the above with respect to q(x) and f(x). Then the asymptotic behavior of the solution of the equations $L_\varepsilon y \equiv \varepsilon y'' + xy' - q(x)y = f(x),\ x \in [0,1]$, and $y(0)=0,\ y(1)=0$, can be written as:

$$\sum_{k=0}^{\infty} \mu^k \left(z_k(x) + \pi_k(t)\right),\ \varepsilon = \mu^2,\ x = \mu t,$$

where $z_k(x) \in C^\infty[0,1], \pi_k(t) \in C^\infty[0,\mu^{-1}]$.

Function $z_0(x)$ is a solution of equation,

$$Mz_0 = f(x) - c_0 xp(x),$$

Where $c_0 = p(0)^{-1}\left[f'(0) - q'(0)f(0)\right]$.

The coefficients $z_k(x)$ of the series $\sum_{k=o}\mu^k\left(z_k(x)+\pi_k(t)\right)$, $\varepsilon=\mu^2$, $x=\mu t$, will be determined as the solution of equations,

$$Mz_k=-z''_{k-1}(x)-c_k xp(x),$$

where $c_k=p(o)^{-1}\left[-z'''_{k-1}(o)+z''_{k-1}(o)q'(o)\right]$, with boundary conditions $z_k(1)=o, k\geq 1$

Functions $\pi_k(t)$ is the solution of the equations,

$$L\pi_k\equiv\pi''_k(t)+t\pi'_k(t)-q(\mu t)\pi_k(t)-c_k\mu tp(\mu t)$$

with boundary conditions $\pi_k(o)=-z_k(o), \pi_k(\mu^{-1})=o.$

Next, we use the following lemma.

Lemma: The problem,

$$My=f\ x)-r\ xp\quad)$$

It has a unique solution $y(x)\in C^\infty[o,1]$.

The proof of above given Lemma follows from equations $y_o(x)=a(x)+r_1 xp(x)\ln x$, and

$$a(x)=xp(x)\int_1^x\left(r(s)-r_o-r_1 s\right)s^{-2}ds+r_o p(x)[x-1],$$
$$r_o=r(o), r_1=r'(o)=p(o)^{-1}\left[f'(o)-q'(o)f(o)\right].$$

Lemma: A boundary value problem,

$$L_o v\equiv v''+tv'-v(t)=o,\ v(o)=a,\ v(1/\mu)=o,$$

has solution $v(t)=aX(t)$, where,

$$X(t)=t\int_t^{\mu^{-1}}s^{-2}\exp\left(\frac{-s^2}{2}\right)ds,\ o\leq X(t)\leq 1,\ X(o)=1.$$

The proof of this Lemma is obvious.

Lemma: In order to solve the boundary value problem,

$$L_o W=-\mu t, W(o)=W(\mu^{-1})=o,$$

we have the estimate

$$o\leq W(\mu,t)\leq e^{-1}\ln\mu^{-1}.$$

Proof: This follows from the fact that the solution of this problem exists uniquely by the maximum principle and will be represented in the form,

$$W(\mu,t)=\mu t\int_t^{\mu^{-1}} y^{-2}\exp\left(\frac{-y^2}{2}\right)\int_0^y s^2\exp\left(\frac{s^2}{2}\right)dsdy.$$

Lemma: The estimate,

$$|\pi_k(\mu,t)|<B_k,$$

Where $0<B_k$ is constant.

Proof: Consider the function,

$$V_\pm(\mu,t)=\gamma_1 W(\mu,t)+\gamma_2 X(t)\pm\pi_k(\mu,t),$$

where γ_1 and γ_2 are positive constants such that,

$$\gamma_1>\max_{[0,1]}|p(x)|,\gamma_2>|z_k(0)|.$$

It is obvious that,

$$V_\pm(\mu,0)>0, V_\pm(\mu,\mu^{-1})>0,\ L_0V_\pm\equiv V''_\pm(t)+tV'_\pm(t)-V_\pm(t)<0$$

From the maximum principle, it follows that,

$$|\pi_k(\mu,t)|<\gamma_1 W(\mu,t)+\gamma_2 X(t).$$

Now the proof of this lemma follows from estimates of $W(\mu,t)$ and $X(t)$.

If we introduce the notation,

$$Y_n(x,\varepsilon)=\sum_{k=0}^{n}\varepsilon^k\left(z_k(x)+\pi_k(\mu,t)\right),$$

where $z_k(x),\pi_k(\mu,t)$ are constructed above functions, then,

$$L_\varepsilon Y_n(x,\varepsilon)=f(x)+\varepsilon^{n+1}z''_n.$$

Let $y(x,\varepsilon)$ be the solution of the equations $L_\varepsilon y\equiv\varepsilon y''+xy'-q(x)y=f(x)$, $x\in[0,1]$, and $y(0)=0$, $y(1)=0$. Then,

$$|L_\varepsilon(Y_n(x,\varepsilon)-y(x,\varepsilon))|<B_n\varepsilon^{n+1}, Y_n(0,\varepsilon)-y(0,\varepsilon)=Y_n(1,\varepsilon)-y(1,\varepsilon)=0.$$

Therefore, $|Y_n(x,\varepsilon)-y(x,\varepsilon)|<B_n\varepsilon^{n+1}$.

The Bisingular Problem of Cole Equation with a Weak Singularity

The following problem is considered,

$$\varepsilon y''(x)+\sqrt{x}y'(x)-y(x)=0, 0<x<1,$$

$$y(0)=a,\ y(1)=b$$

where $x\in[0,1]$; a,b are the given constants.

The unperturbed equation $\sqrt{x}y'(x)-y(x)=0, 0<x<1,$ has the general solution,

$$y_0(x)=ce^{2\sqrt{x}}, c-\text{const.}$$

This is a nonsmooth function.

We seek asymptotic representation of the solution of the equations $\varepsilon y''(x)+\sqrt{x}y'(x)-y(x)=0, 0<x<1,$ and $y(0)=a,\ y(1)=b$ in the form:

$$y(x)=\sum_{k=0}^{n}\varepsilon^k y_k(x)+\sum_{k=0}^{3(n+1)}\mu^k\pi_k(t)+R(x,\varepsilon),$$

where $t=x/\mu^2, \varepsilon=\mu^3, y_k(x)\in C[0,1], \pi_k(t)\in C[0,1/\mu^2], R(x,\varepsilon)$ is the reminder term.

Substituting equation $y(x)=\sum_{k=0}^{n}\varepsilon^k y_k(x)+\sum_{k=0}^{3(n+1)}\mu^k\pi_k(t)+R(x,\varepsilon),$ into equation $\varepsilon y''(x)+\sqrt{x}y'(x)-y(x)=0, 0<x<1,$ we have,

$$\sum_{k=1}^{n}\varepsilon^k\left(\varepsilon y_k''(x)+\sqrt{x}y_k'(x)-yk(x)\right)+\frac{1}{\mu}\left(\pi_0''(t)+\sqrt{t}\pi_0'(t)\right)$$

$$+\sum_{k=1}^{3(n+1)}\mu^{k-1}\left(\pi''_k(t)+\sqrt{t}\pi'k(t)-\pi_{k-1}(t)\right)-\mu^{3(n+1)}\pi_{3(n+1)}(t)+\varepsilon R''(x,\varepsilon)+\sqrt{x}R'(x,\varepsilon)$$

$$-R(x,\varepsilon)-h(x,\varepsilon)+h(x,\varepsilon)=0$$

By the method of generalized boundary layer function, we put the term $h(x,\varepsilon)=\sum_{k=0}^{n-1}\varepsilon^k h_k(x)$ into the equation. We choose functions $h_k(x)$ so that $y_k(x)\in C[0,1]$.

Taking into account the boundary condition of equation $y(0)=a,\ y(1)=b$, from equation,

$$\sum_{k=1}^{n}\varepsilon^k\left(\varepsilon y_k''(x)+\sqrt{x}y_k'(x)-yk(x)\right)+\frac{1}{\mu}\left(\pi_0''(t)+\sqrt{t}\pi_0'(t)\right)$$

$$+\sum_{k=1}^{3(n+1)}\mu^{k-1}\left(\pi''_k(t)+\sqrt{t}\pi'k(t)-\pi_{k-1}(t)\right)-\mu^{3(n+1)}\pi_{3(n+1)}(t)+\varepsilon R''(x,\varepsilon)+\sqrt{x}R'(x,\varepsilon),$$

$$-R(x,\varepsilon)-h(x,\varepsilon)+h(x,\varepsilon)=0$$

we obtain,

$$\sqrt{x}y'o(x)-y_o(x)=0,0<x<1,\quad y_o(1)=b.$$

$$\sqrt{x}y_k'(x)-y_k(x)=h_{k-1}(x)-y''_{k-1}(x),\ 0<x<1,\ k\in N,\ y_k(1)=0.$$

The solution of the equations above exists. It is unique and has the form,

$$y_o(x)=be^{2(\sqrt{x}-1)},y_k(x)=e^{2\sqrt{x}}\int_1^x\frac{h_{k-1}(s)-y_{k-1}''(s)}{\sqrt{s}}e^{-2\sqrt{s}}ds,\ k\in N.$$

We choose indefinite functions $h_k(x)$ as follows:

$y''_{k-1}(x)-h_{k-1}(x)\in C[0,1]$. We can represent:

$$y_o(x)=be^{-2}\left(1+2\sqrt{x}+\frac{(2\sqrt{x})^2}{2!}+\frac{(2\sqrt{x})^3}{3!}+\frac{(2\sqrt{x})^4}{4!}+\ldots+\frac{(2\sqrt{x})^n}{n!}+\ldots\right).$$

Let $h_1(x)=be^{-2}\left(2\sqrt{x}+\frac{(2\sqrt{x})^3}{3!}\right)''=-be^{-2}\left(\frac{1}{2\sqrt{x^3}}-\frac{1}{\sqrt{x}}\right).$

Then,

$$y''_o(x)-h_o(x)\in C[0,1],\mu^3h_1(t\mu^2)=-c_1\left(\frac{1}{2\sqrt{t^3}}-\frac{\mu^2}{\sqrt{t}}\right),c_1=be^{-2},$$

$$y_1(x)=c_1e^{2\sqrt{x}}\int_1^x\left(-\frac{1}{2s^2}+\frac{1}{s}+\frac{1}{2s^2}e^2\sqrt{s}-\frac{1}{\sqrt{s^3}}e^{2\sqrt{s}}\right)e^{-2\sqrt{s}}ds.$$

We can rewrite $y_1(x)$ in the form:

$$y_1(x)=y_{1,0}+y_{1,1}(2\sqrt{x})+y_{1,2}(2\sqrt{x})^2+y_{1,3}(2\sqrt{x})^3+\ldots,$$

where,

$$y_{1,0}=\left(\frac{3}{2}+\frac{1}{2e^2}\right)c_1,y_{1,1}=\left(\frac{1}{6}+\frac{1}{2e^2}\right)c_1,y_{1,2}=\left(\frac{-1}{6}+\frac{1}{4e^2}\right)c_1,y_{1,3}=\left(\frac{-1}{10}+\frac{1}{12e^2}\right)c_1.$$

Analogously, we have obtained,

$$h_1(x)=\left(y_{1,1}(2\sqrt{x})+y_{1,3}(2\sqrt{x})^3\right)''=-\frac{y_{1,1}}{2\sqrt{x^3}}+\frac{6y_{1,3}}{\sqrt{x}}.$$

Then,

$$y''_2(x)-h_2(x)\in C[0,1],\mu^6h_2(t\mu^2)=-\frac{\mu^3y_{1,1}}{2\sqrt{t^3}}+\frac{\mu^5y_{1,3}}{\sqrt{t}}.$$

Continuing this process, we have,

$$h_{k-1}(x)=-\frac{y_{k-1,1}}{2\sqrt{x^3}}+\frac{6y_{k-1,3}}{\sqrt{x}},k=4,\ldots,n,$$

where $y_{k-1,1},y_{k-1,3}$ are corresponding coefficients of the expansion of $y_{k-1,1}(x)$ in powers of $\left(2\sqrt{x}\right)$.

From equation,

$$\sum_{k=1}^{n}\varepsilon^k\left(\varepsilon y_k''(x)+\sqrt{x}y_k'(x)-yk(x)\right)+\frac{1}{\mu}\left(\pi_0''(t)+\sqrt{t}\pi_0'(t)\right)$$

$$+\sum_{k=1}^{3(n+1)}\mu^{k-1}\left(\pi''_k(t)+\sqrt{t}\pi'k(t)-\pi_{k-1}(t)\right)-\mu^{3(n+1)}\pi_{3(n+1)}(t)+\varepsilon R''(x,\varepsilon)+\sqrt{x}R'(x,\varepsilon),$$

$$-R(x,\varepsilon)-h(x,\varepsilon)+h(x,\varepsilon)=0$$

we have the following equations for the boundary functions $\pi_k(t)$:

$$L\pi_0\equiv\pi_0''(t)+\sqrt{t}\pi_0'(t)=0,\ 0<t<\tilde{\mu},\ \pi_0(0)=a-y_0(0),\ \pi_0(\tilde{\mu})=0,\ \tilde{\mu}=1/\mu^2,$$

$$L\pi_{3k+1}(t)=\pi_{3k}(t)+\frac{y_{k,1}}{2\sqrt{t^3}},\ 0<t<\tilde{\mu},\ \pi_{3k+1}(0)=0,\ \pi_{3k+1}(\tilde{\mu})=0,\ k=0,1,\ldots,n$$

$$L\pi_{3k+2}(t)=\pi_{3k+1}(t),\ 0<t<\tilde{\mu},\ \pi_{3k+2}(0)=0,\ \pi_{3k+2}(\tilde{\mu})=0,\ k=0,1,\ldots,n$$

$$L\pi_{3k+3}=\pi_{3k+2}(t)-\frac{y_{k,3}}{\sqrt{t}},\ 0<t<\tilde{\mu},\ \pi_{3k}(0)=-y_k(0),\ \pi_{3k}(\tilde{\mu})=0,\ k=0,1,\ldots,n-1$$

$$L\pi_{3(n+1)}(t)=\pi_{3n+2}(t)-\frac{y_{n,3}}{\sqrt{t}},\ 0<t<\tilde{\mu},\ \pi_{3n}(0)=0,\ \pi_{3n}(\tilde{\mu})=0$$

The solution of equation $L\pi_0\equiv\pi_0''(t)+\sqrt{t}\pi_0'(t)=0,\ 0<t<\tilde{\mu},\ \pi_0(0)=a-y_0(0),\ \pi_0(\tilde{\mu})=0,\ \tilde{\mu}=1/\mu^2$, is represented in the form,

$$\pi_0(t)=(a-be^{-2})A\int_t^{\tilde{\mu}}e^{-\frac{2}{3}s^{3/2}}ds,\ A=\left(\int_t^{\tilde{\mu}}e^{-\frac{2}{3}s^{3/2}}ds\right)^{-1}.$$

We note that $\pi_0(t)$ will exponentially decrease as $t \to \tilde{\mu}$.

Lemma: The general solution of this equation $Lz(t)=0$ will have $z(t)=c_1Y(t)+c_2X(t)$; here $c1, c2$ are constants, and

$$Y(t)=1-X(t),\ X(t)=\alpha\int_t^{\tilde{\mu}} e^{-\frac{2}{3}s^{3/2}}ds\left(\alpha\int_0^{\tilde{\mu}} e^{-\frac{2}{3}s^{3/2}}ds=1\right).$$

Two linearly independent solutions and $Y(t)=O(t), t\to 0, 0<X(t)\le 1$,

$$X(t)=t^{-\frac{1}{2}}e^{-\frac{2}{3}t^{3/2}}\left(1-\frac{1}{2}t^{-\frac{3}{2}}+\ldots+\frac{(-1)^n}{2^n}\prod_{n}^{k=1}1\cdot 4\cdot\ldots\cdot(3k-2)t^{-\frac{3n}{2}}+\ldots\right),\ t\to\tilde{\mu}$$

Lemma: The boundary problem $Lz(t)=0, z(0)=z(\tilde{\mu})=0$ will have only trivial solution.

Theorem: The problem,

$$Lz(t)=f(t),\ z(0)=0,\ z(\tilde{\mu})=0,$$

will have the unique solution and this one has the next form,

$$z(t)=\int_0^{\tilde{\mu}} G(t,s)e^{\frac{2}{3}s^{3/2}}f(s)ds,$$

and $G(t,s)=\begin{cases}-Y(t)X(s), & 0\le t\le s,\\ -Y(s)X(t), & s\le t\le\tilde{\mu},\end{cases}$
is the function of Green and $f(t)\in C(0,\tilde{\mu}]$.

the above given Theorem implies the existence and uniqueness of the solution of equation,

$$L\pi_{3k+1}(t)=\pi_{3k}(t)+\frac{y_{k,1}}{2\sqrt{t^3}},\ 0<t<\tilde{\mu},\ \pi_{3k+1}(0)=0,\ \pi_{3k+1}(\tilde{\mu})=0,\ k=0,1,\ldots,n$$

$$L\pi_{3k+2}(t)=\pi_{3k+1}(t),\ 0<t<\tilde{\mu},\ \pi_{3k+2}(0)=0,\ \pi_{3k+2}(\tilde{\mu})=0,\ k=0,1,\ldots,n$$

$$L\pi_{3k+3}=\pi_{3k+2}(t)-\frac{y_{k,3}}{\sqrt{t}},\ 0<t<\tilde{\mu},\ \pi_{3k}(0)=-y_k(0),\ \pi_{3k}(\tilde{\mu})=0,\ k=0,1,\ldots,n-1$$

$$L\pi_{3(n+1)}(t)=\pi_{3n+2}(t)-\frac{y_{n,3}}{\sqrt{t}},\ 0<t<\tilde{\mu},\ \pi_{3n}(0)=0,\ \pi_{3n}(\tilde{\mu})=0:$$

$|\pi_k(t)|<l=\text{const}, t\in[0,\tilde{}]$.

Lemma: Asymptotical expansions of functions $\pi_k(t), t\to\tilde{\mu}(k=1,2,\ldots)$ will have the next forms,

$$\pi_1(t) = -\frac{y_{0,1}}{2t}\left(1 + \frac{4}{5\sqrt{t^3}} + \frac{7}{4t^3} + \frac{42}{11\sqrt{t^9}} + \frac{39}{2t^7} + \ldots\right),$$

$$\pi_2(t) = -\frac{y_{0,1}}{\sqrt{t}}\left(1 + \frac{23}{40\sqrt{t^3}} + \frac{173}{2t^3} + \ldots\right),\ \pi_3(t) = -\frac{23y_{0,1}}{60\sqrt{t^3}} + O\left(\frac{1}{t^3}\right),$$

Proof for Lemma:

First proof: We can prove this lemma by applying equation,

$$X(t) = t^{-\frac{1}{2}} e^{-\frac{2}{3}t^{3/2}}\left(1 - \frac{1}{2}t^{-\frac{3}{2}} + \ldots + \frac{(-1)^n}{2^n}\prod_{n}^{k=1} 1\cdot 4\cdot\ldots\cdot(3k-2)t^{-\frac{3n}{2}} + \ldots\right), t \to \tilde{\mu}$$ and above given

Theorem.

Second proof: We can receive these representations from equations,

$$L\pi_{3k+1}(t) = \pi_{3k}(t) + \frac{y_{k,1}}{2\sqrt{t^3}},\ 0 < t < \tilde{\mu},\ \pi_{3k+1}(0) = 0,\ \pi_{3k+1}(\tilde{\mu}) = 0,\ k = 0,1,\ldots,n$$

$$L\pi_{3k+2}(t) = \pi_{3k+1}(t),\ 0 < t < \tilde{\mu},\ \pi_{3k+2}(0) = 0,\ \pi_{3k+2}(\tilde{\mu}) = 0,\ k = 0,1,\ldots,n$$

$$L\pi_{3k+3} = \pi_{3k+2}(t) - \frac{y_{k,3}}{\sqrt{t}},\ 0 < t < \tilde{\mu},\ \pi_{3k}(0) = -y_k(0),\ \pi_{3k}(\tilde{\mu}) = 0,\ k = 0,1,\ldots,n-1$$

$$L\pi_{3(n+1)}(t) = \pi_{3n+2}(t) - \frac{y_{n,3}}{\sqrt{t}},\ 0 < t < \tilde{\mu},\ \pi_{3n}(0) = 0,\ \pi_{3n}(\tilde{\mu}) = 0$$

$$|\pi_k(t)| < 1 = \text{const},\ t \in [0, \tilde{\ }].$$

Now we will prove the boundedness of the reminder function $R(x,\varepsilon)$. This function will satisfy the next equation:

$$\varepsilon R''(x,\varepsilon) + \sqrt{x}R'(x,\varepsilon) - R(x,\varepsilon) = \mu^{3(n+1)}\pi_{3(n+1)}(t) + \varepsilon^{n+1}(h_n(x) - y''_n(x)),$$

$$R(0,\varepsilon) = 0,\ R(1,\varepsilon) = 0.$$

Applying to this problem theorem we obtained,

$$|R(x,\varepsilon)| \le \varepsilon^{n+1} C \max_{\substack{0\le x\le 1\\ 0\le t\le\tilde{\mu}}} |\pi_{3(n+1)}(t) + h_n(x) - y''_n(x)|.$$

Therefore, we have $R(x,\varepsilon) = O(\varepsilon^{n+1}), \varepsilon \to 0, x \in [0,1]$.

We prove next.

Theorem: The asymptotical expansion of the solution of the equations $\varepsilon y''(x)+\sqrt{x}y'(x)-y(x)=0, 0<x<1$, and $y(0)=a$, $y(1)=b$ and will have the next form

$$y(x)=\sum_{k=1}^{n}\varepsilon^k yk(x)+\sum_{k=0}^{3(n+1)}\mu^k\pi_0(t)+O\left(\varepsilon^{n+1}\right),\ \varepsilon\to 0.$$

Singularly Perturbed Differential Equations Lighthill Type

The idea of the method of Poincare:

Consider the equation:

$$My(x):=y''(x)+y(x)-\varepsilon y^3(x)=0$$

Unperturbed equation has solutions $y_0(x)=a_1\cos x+b_1\sin x$ (where a_1, b_1 are arbitrary constants) with period 2π. We are looking for the periodic solution of the equation $y(x,\varepsilon)$ with a period of $\omega(\varepsilon)=\omega(0)=2\pi$.

Note that the operator M transforms Fourier series $\sum_{1}^{\infty}a_k\cos kx$ and $\sum_{k=1}^{\infty}a_k\cos kx$ in itself. Poincare's method reduces the existence of periodic solutions of differential equations to the existence of the solution of an algebraic equation.

We will seek a periodic solution of equation $My(x):=y''(x)+y(x)-\varepsilon y^3(x)=0$ with the initial condition,

$$y(0)=1,\ y'(0)=0.$$

If we seek the solution in the form,

$$y(x)=y_0(x)+\varepsilon y_1(x)+\varepsilon^2 y_2(x)+\ldots$$

with the initial conditions,

$$y_0(0)=1,\ y'_0(1)=0,\ y_k(0)=y'k(1)=0,\ k=1,2,\ldots$$

then for $y_s(x), s=0,1,\ldots$ we have next equations,

$$Ly_0:=y''_0(x)+y_0(x)=0\Rightarrow y_0(x)=\cos x$$

$$Ly_1=\cos^3 x=\frac{3}{4}\cos x+\frac{1}{4}\cos 3x\Rightarrow y_1(x)=\frac{3}{8}x\sin x-\frac{1}{32}\cos 3x+\frac{1}{32}\cos x,$$

Thus, $y(x)=\cos x+\frac{\varepsilon}{8}(3x\sin x-\frac{1}{4}\cos 3x+\frac{1}{4}\cos x)+\ldots$ it is not a uniform expansion of the $y(x)$ on the segment $[-\infty,\infty]$, since the term $\varepsilon x\sin x$ is present here.

If these secular terms do not appear in equation $My(x) := y''(x) + y(x) - \varepsilon y^3(x) = 0$, it is necessary to make the substitution,

$$x = t(1 + \varepsilon\alpha_1 + \varepsilon^2\alpha_2 + \ldots)$$

where the constant α_k should be selected so as not to have secular terms in t.

Thus, the solution of equation $My(x) := y''(x) + y(x) - \varepsilon y^3(x) = 0$ must be sought in the form,

$$y(t) = y_0(t) + \varepsilon y_1(t) + \varepsilon^2 y_2(t) + \ldots$$
$$x = t(1 + \varepsilon\alpha^1 + \varepsilon^2\alpha_2 + \ldots)$$

Then equation $My(x) := y''(x) + y(x) - \varepsilon y^3(x) = 0$ has the form,

$$z''(t) + (1 + \alpha_1\varepsilon + \alpha_2\varepsilon^2 + \ldots)z(t) = \varepsilon(1 + \alpha_1\varepsilon + \alpha_2\varepsilon^2 + \ldots)z^3(t)$$

where $y(w(\varepsilon)t) = z(t)$.

We will seek the 2π periodic solution of this equation in the form,

$$z(t) = z_0(t) + \varepsilon z_1(t) + \varepsilon^2 z_2(t) + \ldots$$

Then,

$$Lz_0 := z''_0(x) + z_0(x) = 0 \Rightarrow z_0(t) = \cos t.$$

$$Lz_1 = \alpha_1 \cos t + \frac{3}{4}\cos t + \frac{1}{4}\cos 3t.$$

The function $Z_1(t)$ will have the periodical solution we take $\alpha_1 = -3/4$. Then $z_1(t) = -\frac{1}{32}\cos 3t.$

Similarly, from equations,

$$\alpha z_n(t) = -\alpha_n \cos t + g(\alpha_1, \alpha_2, \ldots, \alpha_{n-1})\cos t + \sum_{m=1}^{2n+1} \beta_n \cos mt$$

α_n and etc. are uniquely determined.

Theorem: Equation $My(x) := y''(x) + y(x) - \varepsilon y^3(x) = 0$ has a unique $2\pi/\omega$ periodic solution, and it can be represented in the form of equation $\begin{matrix} y(t) = y_0(t) + \varepsilon y_1(t) + \varepsilon^2 y_2(t) + \ldots \\ x = t(1 + \varepsilon\alpha^1 + \varepsilon^2\alpha_2 + \ldots) \end{matrix}$.

The Idea of the Lighthill Method

Lighthill in 1949 reported an important generalization of the method of Poincare.

He considered the model equation:

$$(x+\varepsilon y(x))y'(x)+q(x)y(x)=r(x), \quad y(1)=a$$

where $x\in[0,1]$ $q(x), r(x)\in C^{\infty}[0,1]$.

Lighthill proposed to seek the solution of equation above in the form,

$$y(\xi)=y_0(\xi)+\varepsilon y_1(\xi)+\varepsilon^2 y_2(\xi)+\ldots$$
$$x=\xi+\varepsilon x_1(\xi)+\varepsilon^2 x_2(\xi)+\ldots$$

It is obvious that equation above has generalized the Poincare ideas.

At first, we consider the example:

$$(x+\varepsilon y(x))y'(x)+y(x)=0, \quad y(1)=b.$$

It has exact solution,

$$y(x)=\left(\sqrt{x^2+2b\varepsilon+\varepsilon^2 b^2}-x\right)/\varepsilon.$$

It is obvious that for $b>0$, the solution of equation $(x+\varepsilon y(x))y'(x)+y(x)=0, \quad y(1)=b.$ exists on the interval [0, 1] and

$$y(0)=\sqrt{2b+\varepsilon^2}/\sqrt{\varepsilon}.$$

The solution of equation $(x+\varepsilon y(x))y'(x)+y(x)=0, \quad y(1)=b$ is obtained by the method of small parameter that can be obtained from equation $y(x)=\left(\sqrt{x^2+2b\varepsilon+\varepsilon^2 b^2}-x\right)/\varepsilon$.

For this purpose, we write equation $y(x)=\left(\sqrt{x^2+2b\varepsilon+\varepsilon^2 b^2}-x\right)/\varepsilon$ in the form,

$$y(x)=\frac{x}{\varepsilon}\left(-1+\sqrt{1+2b\frac{\varepsilon}{x}+b^2\left(\frac{\varepsilon}{x}\right)^2}\right)$$

and considering $x^2>2\varepsilon b$, this expression can be expanded in powers of ε, and then we have,

$$y(x)=\frac{b}{x}+\frac{b^2}{2x}\frac{\varepsilon}{x^2}(x^2-1)+\ldots+O\left(\frac{1}{x}\left(\frac{\varepsilon}{x^2}\right)^n\right)+\ldots$$

The series is uniformly convergent asymptotic series only on the segment $[\varepsilon^{\alpha},1], 0<\alpha<1/2$.

First, we write equation $(x+\varepsilon y(x))y'(x)+y(x)=0,\ \ y(1)=b$ in the form,

$$(x+\varepsilon y(\xi))y'(\xi)+y(\xi)x'(\xi)=0$$

Substituting equation $\begin{matrix} y(\xi)=y_0(\xi)+\varepsilon y_1(\xi)+\varepsilon^2 y_2(\xi)+\dots \\ x=\xi+\varepsilon x_1(\xi)+\varepsilon^2 x_2(\xi)+\dots \end{matrix}$ into equation $(x+\varepsilon y(\xi))y'(\xi)+$ $y(\xi)x'(\xi)=0$:

$$\left(\xi+\varepsilon\left(y_0(\xi)+x_1(\xi)\right)+\dots+\varepsilon^n\left(y_{n-1}(\xi)+x_n(\xi)\right)+\dots\right)\left(y'_0(\xi)+\varepsilon y'1(\xi)+\dots++\varepsilon^n y'_n(\xi)+\dots\right)$$
$$+\left(y_0(\xi)+\varepsilon y'_1(\xi)+\dots\varepsilon^n y'_n(\xi)+\dots\right)\left(1+\varepsilon x'_1(\xi)+\dots+\varepsilon^n x'_n(\xi)+\dots\right)=0$$

and equating coefficients of the same powers ε, we have,

$$\xi y'_0(\xi)+y_0(\xi)=0$$

$$\xi y'_n(\xi)+y_n(\xi)+\sum_{i=0}^{n-1}\left((y_i(\xi)+x_{i+1}(\xi))y'_{n-1-i}(\xi)+y_i(\xi)x'_{n-i}(\xi)\right)=0,\quad y_n(1)=0, n=1,2,\dots$$

From equation $\xi y'_0(\xi)+y_0(\xi)=0$ we have,

$$y_0(\xi)=b\xi^{-1}.$$

Using equation $\xi y'_0(\xi)+y_0(\xi)=0$, equation $\xi y'_n(\xi)+y_n(\xi)+\sum_{i=0}^{n-1}((y_i(\xi)+x_{i+1}(\xi))y'_{n-1-i}$ $(\xi)+y_i(\xi)x'_{n-i}(\xi))=0,\ \ y_n(1)=0, n=1,2,\dots$ for $n=1$ can be written as,

$$\xi y'_1(\xi)+y_1(\xi)=(\xi x'_1(\xi)-x_1(\xi)+y_0(\xi))y'_0(\xi)=0,\quad y_1(1)=0.$$

If we put $x_1(\xi)=0$ in the above equation we obtain,

$$\xi y'_1(\xi)+y_1(\xi)=-b^2\xi^{-3},\quad y_1(1)=0.$$

Hence, solving this equation, we have,

$$y_1(\xi)=b^2(2\xi)^{-1}-b^2(2\xi^3)^{-1}.$$

Since differentiation increased singularity of nonsmooth function, we select $x_1(\xi)$ so that the expression in the right side of equation $\xi y'_1(\xi)+y_1(\xi)=(\xi x'_1(\xi)-x_1(\xi)+y_0(\xi))y'_0(\xi)=0,\ \ y_1(1)=0$ is equal to zero, i.e.,

$$\xi x'_1(\xi)-x_1(\xi)+y_0(\xi)=0,\quad x_1(1)=0.$$

Hence, we have,

$$x_1(\xi)=2^{-1}b\xi-(2\xi)^{-1}b.$$

Then equation $\xi y'_1(\xi)+y_1(\xi)=(\xi x'_1(\xi)-x_1(\xi)+y_0(\xi))y'_0(\xi)=0,\quad y_1(1)=0$ takes the form,

$$\xi y'_1(\xi)+y_1(\xi)=0,\quad y_1(1)=0.$$

Hence, we obtain $y_1(\xi)=0$.

Now equation,

$$\xi y'_n(\xi)+y_n(\xi)+\sum_{i=0}^{n-1}((y_i(\xi)+x_{i+1}(\xi))y'_{n-1-i}(\xi)+y_i(\xi)x'_{n-i}(\xi))=0,\quad y_n(1)=0, n=1,2,\ldots \text{ for n=2}$$

takes the form,

$$\xi y'_2(\xi)+y_2(\xi)=(\xi x'_2(\xi)-x_2(\xi))y'_0(\xi)=0,\quad y_2(1)=0.$$

Let $x_2(\xi)=0$, and then $y_1(\xi)=0$. Further also choose $x_i(\xi)=y_i(\xi)=0\ (i=3,4,\ldots)$. as they also satisfy the initial conditions. Thus, we have found that,

$$y(\xi)=b\xi^{-1}$$

$$x(\xi)=\xi+\frac{b}{2}\left(\xi-\frac{1}{\xi}\right)\varepsilon.$$

Putting in equation above $x=0$, we have,

$$\eta=\sqrt{b\varepsilon/(2+b\varepsilon)}.$$

For $b>0$, the point $x=0$ is achieved. Moreover, the except in variable ξ from equation $y(\xi)=b\xi^{-1}$ and to equation $x(\xi)=\xi+\frac{b}{2}\left(\xi-\frac{1}{\xi}\right)\varepsilon$ setting ξ, we obtain the exact solution of equation $y(x)=\left(\sqrt{x^2+2b\varepsilon+\varepsilon^2b^2}-x\right)/\varepsilon$.

Now we will present the main idea of the Lighthill method to equation $(x+\varepsilon y(x))y'(x)+q(x)y(x)=r(x),\quad y(1)=a$ under conditions: $q(x),r(x)\in C^\infty[0,1]$ and $q_0=q(0)>0$. We will write it in the form of,

$$(x(\xi)+\varepsilon y(\xi))y'(\xi)=[r(x(\xi))-q(x(\xi))y(\xi)]x'(\xi),\quad y(1)=y^0.$$

It is obvious that we have one equation for two unknown functions, $y(\xi),x(\xi)$. Now we substitute the series of equation $\begin{matrix}y(\xi)=y_0(\xi)+\varepsilon y_1(\xi)+\varepsilon^2y_2(\xi)+\ldots\\ x=\xi+\varepsilon x_1(\xi)+\varepsilon^2x_2(\xi)+\ldots\end{matrix}$ to the above equation.

$$\left(\xi+\sum_{k=0}^{\infty}\varepsilon^k\left(y_k(\xi)+x_k(\xi)\right)\right)\sum_{k=0}^{\infty}\varepsilon^k y'_k(\xi)$$

$$=\left(\sum_{j=0}^{\infty}r_j(\xi)\left(\sum_{k=0}^{\infty}x_k(\xi)\varepsilon^k\right)^j-\sum_{j=0}^{\infty}q_j(\xi)\left(\sum_{k=0}^{\infty}x_k(\xi)\varepsilon^k\right)^j\right)\left(1+\sum_{k=0}^{\infty}x'_k(\xi)\varepsilon^k\right),$$

where $q_j=q_j(\xi)=\frac{1}{j!}q^{(j)}(\xi)$, $r_j=r_j(\xi)=\frac{1}{j!}r^{(j)}(\xi)$.

Hence, equating the coefficients of equal powers has ε,

$$Lu_o\equiv\xi y'_o(\xi)+q(\xi)y_o(\xi)=r(\xi),\ y_o(1)=y^o,$$

$$Ly_1=[\xi y'_o x'_1-y'_o x_1-y_o y'_o]+(r_1-q_1y_o)x_1,\ y_1(1)=0,$$

$$Ly_2=\left[\xi y'_o x'_2-(y_o+x_1)y'_1-(y_1+x_2)y'_o+((r_1-q_1y_o)x_1-qy_1)x'_1\right]+$$
$$+\left\{r_1x_2+r_2x_1^2-q_1x_1y_1-(q_1x_2+q_2x_1^2)y_o\right\},\ y_2(1)=0,$$

$$Ly_n=\left[y'_o x'_n-y'_o x_n+f_n(y_o,\ldots,y_{n-1},x_1,\ldots,x_{n-1},y'_o,\ldots,y'_{n-1},x'_1,\ldots,x'_{n-1})\right]+$$
$$+\left\{g_n(y_o,\ldots,y_{n-1},x_1,\ldots,x_{n-1})\right\},y_n(1)=0;$$

where $q=q_o$, $r=r_o$,

$$f_n=-(y_o+x_1)y'_{n-1}-(y_1+x_2)y'_{n-2}-\ldots-(y_{n-2}+x_{n-1})y'_1-y_{n-1}y'_o+$$
$$+(r_1x_1-qy_1-q_1x_1y_o)x'_{n-1}+\left(r_1x_2+r_2x_1^2-qy_2-q_1x_1y_1-\left(q_1x_2+q_2x_1^2\right)y_o\right)x'_{n-2}+\ldots$$
$$+\left(r_1x_{n-1}+2_{r2}x_1x_{n-2}+2_{r2}x_2x_{n-3}+\ldots+r_{n-1}x_1^{n-1}-q_1y_1x_{n-2}-\left(q_1x_{n-1}+2_{q2}x_1x_{n-2}+\ldots+\right.\right.$$
$$\left.\left.q_{n-1}x_1^{n-1}\right)y'_o\right)x'1,$$

$$g_n=r_1x_n+2_{r2}x_1x_{n-1}+\ldots+r_nx_1^n-q_1x_1y_{n-1}-\left(q_1x_2+q_2x_1^2\right)y_{n-2}-\ldots$$
$$-\left(q_1x_n+2q_2x_1x_{n-1}+\ldots+q_nx_1^n\right)y_o.$$

In these equations, the coefficient $r(\xi)-q(\xi)y_o(\xi)$ of the derivative $x'_n(\xi)(n=1,2,\ldots)$ was replaced by equation $Lu_o\equiv\xi y'_o(\xi)+q(\xi)y_o(\xi)=r(\xi)$, $y_o(1)=y^o$, on $\xi y'_o(\xi)$.

From equation $Ly_n=\left[y'_o x'_n-y'_o x_n+f_n(y_o,\ldots,y_{n-1},x_1,\ldots,x_{n-1},y'_o,\ldots,y'_{n-1},x'_1,\ldots,x'_{n-1})\right]$ ...$y_n(1)=0$; for $n=1,2,\ldots$, it follows that if we want to define functions $x_n(\xi)$ $(n=1,2,\ldots)$ from this differential equations, then we must assume that,

$$\xi y'_o(\xi)=r(\xi)-q(\xi)y_o(\xi)\neq 0,\ \xi\in(0,1].$$

And this condition cannot be avoided by applying the Lighthill method to equation $(x+\varepsilon y(x))y'(x)+q(x)y(x)=r(x),\ y(1)=a$. Justifying Lighthill method, then in the works Habets, Sibuya, and Takahashi. Comstock on the example shows that the equation $\xi y'_0(\xi)=r(\xi)-q(\xi)y_0(\xi)\neq 0,\ \xi\in(0,1]$ is not necessary for the existence of solutions on the interval $[0,1]$. Further assume that the equation $\xi y'_0(\xi)=r(\xi)-q(\xi)y_0(\xi)\neq 0,\ \xi\in(0,1]$ holds. The right-handside of equation $Ly_n=\left[y'_0x'_n-y'_0x_n+f_n(y_0,\ldots,y_{n-1},x_1,\ldots,x_{n-1},y'_0,\ldots,y'_{n-1},x'_1,\ldots,x'_{n-1})\right]\ldots y_n(1)=0;$ is linear with respect to $x_n(\xi)$, and f_n function depends from $y'_0,\ldots,y'_{n-1},x'_1,\ldots,x'_{n-1}$ only.

The solution of equation $Lu_0\equiv\xi y'_0(\xi)+q(\xi)y_0(\xi)=r(\xi),\ y_0(1)=y^0$, can be written as,

$$y_0(\xi)=\xi^{-q_0}g(\xi)(y^0+\int_1^{\xi}s^{q_0-1}r(s)g^{-1}(s)ds):=\xi^{-q_0}w(\xi),$$

where $g(\xi)=\exp\left(\int_1^{\xi}(q_0-q(s))s^{-1}ds\right)$.

Let,

$$w_0=y^0-\int_0^1 s^{q_0-1}r(s)g^{-1}(s)ds\neq 0\Leftrightarrow w_0=w(0)\neq 0.$$

Hence, we have,

$$y_0(\xi)\sim\xi^{-q_0}w_0,\quad \xi\to 0.$$

Since the differentiation of $y_0(\xi)$ increased of its singularity at the point $\xi=0$, it is better to choose such that the first brace in equation $Ly_1=[\xi y'_0x'_1-y'_0x_1-y_0y'_0]+(r_1-q_1y_0)x_1,\ y_1(1)=0$, is equal to zero, i.e.,

$$\xi x'_1=x_1+y_0,\quad x(1)=0.$$

Hence, using equation $y_0(\xi)\sim\xi^{-q_0}w_0,\ \xi\to 0$ we obtain,

$$x_1(\xi)=\xi+\xi\int_1^{\xi}s^{-2}y_0(s)ds\sim-\frac{w_0}{1+q_0}\xi^{-q_0}.$$

Then equation $Ly_1=[\xi y'_0x'_1-y'_0x_1-y_0y'_0]+(r_1-q_1y_0)x_1,\ y_1(1)=0$, takes the form

$$Ly_1=(r_1-q_1y_0)x_1\sim\tilde{a}_1\xi^{-2q_0},$$

where $\tilde{a}_1=\text{const}$. Hence, we have,

$$y_1(\xi)\sim a_1\,\xi^{-2q_0}\,(a_1=\text{const}),\xi\to 0.$$

Now equating to zero the expression in the first brace in the right-hand side of equation $Ly_2=\left[\xi y'_0x'_2-(y_0+x_1)y'_1-(y_1+x_2)y'_0+((r_1-q_1y_0)x_1-qy_1)x'_1\right]\ldots y_2(1)=0$, we have, $\xi x'_2-x_2=y_1+((y_0+x_1)y'_1-((r_1-q_1y_0)x_1-qy_1)x'_1)(y'_0)^{-1}\sim\tilde{b}_2\xi^{-2q_0},\ \tilde{b}_2=\text{const}$.

From this, we get,

$$x_2(\xi) \sim b_2\xi^{-2q_0},\ b_2 = \text{const},\ \xi \to 0.$$

Now equation $Ly_2 = \left[\xi y'_0 x'_2 - (y_0 + x_1)y'_1 - (y_1 + x_2)y'_0 + ((r_1 - q_1 y_0)x_1 - qy_1)x'_1\right] \ldots y_2(1) = 0$, takes the form:

$$Ly_2 = g_2(y_0, y_1, x_1, x_2) \sim \tilde{a}_2\xi^{-3q_0},\ \tilde{a}_2 = \text{const},\ \xi \to 0$$

Solving this equation, we have,

$$y_2(\xi) \sim a_2\xi^{-3q_0},\ a_2 = \text{const},\ \xi \to 0$$

Next, the method of induction, it is easy to show that,

$$x_j(\xi) \sim b_j\xi^{-jq_0},\ y_j(\xi) \sim a_j\xi^{-(j+1)q_0},\ j = 1, 2, \ldots.$$

Thus, the equation $\begin{matrix} y(\xi) = y_0(\xi) + \varepsilon y_1(\xi) + \varepsilon^2 y_2(\xi) + \ldots \\ x = \xi + \varepsilon x_1(\xi) + \varepsilon^2 x_2(\xi) + \ldots \end{matrix}$ has the asymptotic,

$$y(\xi) \sim \xi^{-q_0}(w_0 + a_1\varepsilon\xi^{-q_0} + \ldots + a_n(\varepsilon\xi^{-q_0})^n + \ldots),\ \xi \to 0,$$

$$x \sim \xi - \frac{w_0}{1+q_0}\xi^{-q_0}\varepsilon + b_2(\varepsilon\xi^{-q_0})^2 + \ldots + b_n(\varepsilon\xi^{-q_0})^n + \ldots$$

From equation above it follows that the point x=0 corresponds to the root of the equation,

$$\eta + \varepsilon x_1(\eta) + \varepsilon^2 x_2(\eta) + \ldots = 0$$

Moreover, this equation should have a positive root and if the solution of equation $(x + \varepsilon y(x))y'(x) + q(x)y(x) = r(x),\ \ y(1) = a$ exists on the interval (0, 1]. Solving the above equation we obtain,

$$\eta \sim (w_0\varepsilon / 1 + q_0)^{1/(1+q_0)},\ \varepsilon \to 0.$$

And, under the condition $w_0 > 0, \eta_0$ will be positive. It is obvious that on the interval

$[\xi_0, 1]$ equations $\begin{matrix} y(\xi) = y_0(\xi) + \varepsilon y_1(\xi) + \varepsilon^2 y_2(\xi) + \ldots \\ x = \xi + \varepsilon x_1(\xi) + \varepsilon^2 x_2(\xi) + \ldots \end{matrix}$ or $y(\xi) \sim \xi^{-q_0}(w_0 + a_1\varepsilon\xi^{-q_0} + \ldots + a_n$ $(\varepsilon\xi^{-q_0})^n + \ldots),\ \xi \to 0,$

and

$x \sim \xi - \frac{w_0}{1+q_0}\xi^{-q_0}\varepsilon + b_2(\varepsilon\xi^{-q_0})^2 + \ldots + b_n(\varepsilon\xi^{-q_0})^n + \ldots$ remains asymptotic. Substituting equation $\eta \sim (w_0\varepsilon/1+q_0)^{1/(1+q_0)}$, $\varepsilon \to 0$ into equation $y(\xi) \sim \xi^{-q_0}(w_0 + a_1\varepsilon\xi^{-q_0} + \ldots + a_n$ $(\varepsilon\xi^{-q_0})^n + \ldots)$, $\xi \to 0$, we have,

$$y(0) \sim w_0\left(\frac{w_0\varepsilon}{1+q_0}\right)^{-q_0/(1+q_0)}, \ \varepsilon \to 0.$$

If $w_0 < 0$ the point $x = 0$ does not have the positive root of equation $\eta + \varepsilon x_1(\eta) + \varepsilon^2 x_2(\eta) + \ldots = 0$, so that the solution of equation $(x + \varepsilon y(x))y'(x) + q(x)y(x) = r(x)$, $y(1) = a$ goes to infinity, before reaching the point $x = 0$.

We have the following:

Theorem: Suppose that the conditions (1) $q(x), r(x) \in C^\infty[0,1]$; (2) $q_0 > 0$; (3) $w_0 > 0$; (4) $\xi y'_0 \neq 0, \xi \in [0,1]$. Then the solution of equation $(x + \varepsilon y(x))y'(x) + q(x)y(x) = r(x)$, $y(1) = a$ exists on the interval [0, 1], and it can be represented in the asymptotic equations $\begin{matrix} y(\xi) = y_0(\xi) + \varepsilon y_1(\xi) + \varepsilon^2 y_2(\xi) + \ldots \\ x = \xi + \varepsilon x_1(\xi) + \varepsilon^2 x_2(\xi) + \ldots \end{matrix}$,

$y(\xi) \sim \xi^{-q_0}(w_0 + a_1\varepsilon\xi^{-q_0} + \ldots + a_n(\varepsilon\xi^{-q_0})^n + \ldots)$, $\xi \to 0$, and $x \sim \xi - \frac{w_0}{1+q_0}\xi^{-q_0}\varepsilon + b_2(\varepsilon\xi^{-q_0})^2 + \ldots + b_n(\varepsilon\xi^{-q_0})^n + \ldots$

Theorem proved by Wasow, Sibuya and Takahashi in the case where $q(x), r(x)$ are analytic functions on $[0,1]$; proved by Habets in the case $q(x), r(x) \in C^2[0,1]$. Moreover, instead of the equation $\varepsilon y'(x) + xy(x) = f(x)$, $0 < x \leq 1$, $y(0) = a$, Wasow impose a stronger condition: $a >> 1$.

In the proof of the above Theorem we will not stop because it is held by Majorant method.

From the foregoing, it follows that Wasow condition $y'_0(\xi) \neq 0$, $\xi \in (0,1]$ is essential in the Lighthill method.

Prytula and later Martin proposed the following variant of the Lighthill method. At first direct expansion determined using by the method of small parameter,

$$y(x) = y_0(x) + \varepsilon y_1(x) + \varepsilon^2 y_2(x) + \ldots$$

and further at second they will make transformation,

$$x = \xi + \varepsilon x_1(\xi) + \varepsilon^2 x_2(\xi) + \ldots$$

Here unknowns $x_j(\xi)$ are determined from the condition that function $y_j(\xi)$ was less singular function $y_{j-1}(\xi)$. We show that using the method Prytula or Martin, also cannot avoid Wasow conditions. Really, substituting equation $x = \xi + \varepsilon x_1(\xi) + \varepsilon\ x_2(\xi) + \ldots$ into equation $y(x) = y_0(x) + \varepsilon y_1(x) + \varepsilon^2 y_2(x) + \ldots$ and expanding in a Taylor series in powers of ε, we have,

$$y(\xi) = y_0(\xi) + \varepsilon\{y_1(\xi) + y'_0(\xi)x_1(\xi)\} + O(\varepsilon^2).$$

Hence, to obtain a uniform representation of the solution to the second order by ε, we must to put to zero the expression in the curly brackets, i.e., $x_1(\xi) = -y_1(\xi)/y'_0(\xi)$. Therefore, $y(\xi) = y_0(\xi) + O(\varepsilon^2)$. Hence, it is clear that we must make the condition of Wasow: $y'_0(\xi) \neq 0$ in the method of Prytula or Martin also.

Uniformization Method for a Lighthill Model Equation

We will consider the equation $(x + \varepsilon y(x))y'(x) + q(x)y(x) = r(x), \quad y(1) = a$ again i.e.,

$$(x + \varepsilon y(x))y'(x) = r(x) - q(x)y(x), \ y(1) = a,$$

Theorem: Suppose that the equation above has a parametric representation of the solution $y = y(\xi), x = x(\xi)$, where $\xi \in [\eta, 1], \eta = \eta(\varepsilon) > 0$, then the equation $(x + \varepsilon y(x))y'(x) = r(x) - q(x)y(x), \ y(1) = a$, is equivalent to the problem,

$$\begin{cases} \xi y'(\xi) = r(x(\xi)) - q(x(\xi))y(\xi), \ y(1) = y^0, \\ \xi x'(\xi) = x(\xi) + \varepsilon y(\xi), x(1) = 1, \quad \xi \in [\eta, 1], \end{cases}$$

where $\eta = \eta(\varepsilon)$ is the root equation $x(\eta) = 0$ and if the root $\eta = \eta(\varepsilon) > 0$ and $x(\xi) + \varepsilon y(\xi) \neq 0$ on the interval $[\eta, 1]$.

Proof – Sufficiency: Let the solution of the equation $(x + \varepsilon y(x))y'(x) = r(x) - q(x)y(x)$, $y(1) = a$, exists and $x(\xi)$, $y(\xi)$ are a parametric representation of the solution of the equation $(x + \varepsilon y(x))y'(x) = r(x) - q(x)y(x), \ y(1) = a$. Then introducing the variable-parameter ξ, we obtain the equation $\{\xi y'(\xi) = r(x(\xi)) - q(x(\xi))y(\xi), \ldots \xi \in [\eta, 1]$.

Necessity: Let it fulfill the conditions of Theorem 8. Then dividing the first equation by second one, we get equation $(x + \varepsilon y(x))y'(x) = r(x) - q(x)y(x), \ y(1) = a$. Theorem above is proved.

Equation $\{\xi y'(\xi) = r(x(\xi)) - q(x(\xi))y(\xi), \ldots \xi \in [\eta, 1]$, on the proposal of the Temple, we will call uniformizing equation for the equation $(x + \varepsilon y(x))y'(x) = r(x) - q(x)y(x), \ y(1) = a$.

We have the following:

Theorem: Suppose that the first three conditions of Theorem a. i.e.,(1) $q(x), r(x) \in C^{\infty}[0,1]$; (2) $q_0 > 0$; (3) $w_0 > 0$. Then the solution of equation $(x+\varepsilon y(x))y'(x) = r(x) - q(x)y(x),\ y(1) = a$ is represented in the form of an asymptotic equations $\begin{matrix} y(\xi) = y_0(\xi) + \varepsilon y_1(\xi) + \varepsilon^2 y_2(\xi) + \dots \\ x = \xi + \varepsilon x_1(\xi) + \varepsilon^2 x_2(\xi) + \dots \end{matrix}$ and its solution can be obtained from uniformizing equation $\{\xi y'(\xi) = r(x(\xi)) - q(x(\xi))y(\xi), \dots \xi \in [\eta, 1]$.

The proof of this theorem is completely analogous to the proof of above given Theorem, even more easily.

Only it remains to show that under the conditions of above given Theorem we can get an explicit solution $y = y(x,\varepsilon)$. Really, since,

$$x \sim \xi - \frac{w_0}{1+q_0}\xi^{-q_0}\varepsilon, \xi \to 0.$$

Let,

$$F(x,\xi,\varepsilon) = x - \xi + \frac{w_0}{1+q_0}\xi^{-q_0}\varepsilon + O\left(\left(\varepsilon^{\xi - q_0}\right)^2\right), \xi \to 0, \eta = \sqrt[q_0+1]{\frac{w_0}{1+q_0}}\varepsilon, \varepsilon \to 0.$$

Then,

$$\frac{\partial F(x,\xi,\varepsilon)}{\partial \xi}|\xi = \eta(\varepsilon) = -1 - q_0 + O(\varepsilon^{1/(1+q_0)}) \neq 0,\ \xi \in [\eta, 1].$$

Therefore, by the implicit function theorem, we can express $\xi : \xi = \phi(x,\varepsilon)$.

Then when we put it in first equality of equations $\begin{matrix} y(\xi) = y_0(\xi) + \varepsilon y_1(\xi) + \varepsilon^2 y_2(\xi) + \dots \\ x = \xi + \varepsilon x_1(\xi) + \varepsilon^2 x_2(\xi) + \dots \end{matrix}$ we obtain an explicit solution $y = y(x,\varepsilon)$.

Example: Uniformized equation is,

$$\begin{cases} \xi y'(\xi) = -y(\xi), & y(1) = b, \\ \xi x'(\xi) = x(\xi) + \varepsilon y(\xi), & x(1) = 1, \quad \xi \in [\eta, 1], \end{cases}$$

It is easy to integrate this system, and we obtain,

$$y(\xi) = b\xi^{-1}, x(\xi) = (1 + 2^{-1}b\varepsilon)\xi - (2\xi)^{-1}b\varepsilon,$$

Hence, excluding variable ξ, we have an exact solution of equation $y(x) = \left(\sqrt{x^2 + 2b\varepsilon + \varepsilon^2 b^2} - x\right)/\varepsilon.$

Example:

$$(x+\varepsilon y(x))y'(x)+(2+x)y(x)=0,\ \ y(1)=e^{-1}.$$

Uniformized equation is,

$$\begin{cases}\xi x'(\xi)=x+\varepsilon y(\xi), & x(1)=1,\\ \xi y'(\xi)=-(2+x(\xi))y(\xi), & y(1)=e^{-1},\quad \xi\in[\eta,1],\end{cases}$$

Let,

$$\begin{cases}x(\xi)=x_0(\xi)+\varepsilon x_1(\xi)+O(\varepsilon^2),\\ y(\xi)=y_0(\xi)+\varepsilon y_1(\xi)+O(\varepsilon^2),\end{cases}$$

Substituting equation above into equation $\{\xi x'(\xi)=x+\varepsilon y(\xi), \ldots y(1)=e^{-1},\ \ \xi\in[\eta,1],$ we have,

$$x_0(\xi)=\xi,\ x_1(\xi)=\xi\int_1^{\xi}e^{-s}s^{-4}ds,\ y_0(\xi)=e^{-\xi}\xi^{-2},\ y_1(\xi)=-e^{-\xi}\xi^{-2}\int_1^{\xi}e^{-s}s^{-4}ds,$$

Hence if $\xi\to 0$, we obtain,

$$x_0(\xi)=\xi,\ x_1(\xi)=-\frac{1}{3}\xi^{-2}+\ldots,\ y_0(\xi)=\xi^{-2}+\ldots,\ y_1(\xi)=-\frac{1}{6}\xi^{-4}+\ldots$$

From the equation $x(\eta)=0$, we find $\eta:\ \eta\sim\sqrt[3]{\varepsilon/3}$.

We prove that $x(\xi)+\varepsilon y(\xi)\neq 0$ on the interval $[\eta, 1]$.

Really,

$$x(\xi)+\varepsilon y(\xi)\sim\xi+\varepsilon\xi^{-2}\neq 0,\xi\in[\eta,1].$$

It is construction Explicit Form of the Solution of the Model Lighthill Equation

We will consider the equation, i.e., equation $(x+\varepsilon y(x))y'(x)+q(x)y(x)=r(x),\ \ y(1)=a$ again,

$$(x+\varepsilon y(x))y'(x)+q(x)y(x)=r(x),\ y(1)=b$$

where b is given constant, $x\in[0,1],\ \ y'(x)=dy/dx$. Given functions are subjected to the conditions $U: q(x), r(x)\in C^{(\infty)}[0,1]$.

Here, we consider the case $q_0=-1$; this is done to provide a detailed illustration of the idea of the application of the method. We search for the solution of equation $(x+\varepsilon y(x))y'(x)+q(x)y(x)=r(x),\ y(1)=b$ in the form,

$$y(x)=\mu^{-1}\pi_{-1}(t)+\sum_{k=o}^{\infty}\left(\pi_k(t)+u_k(x)\right)\mu^k,$$

where $t=x/\mu$, $\varepsilon=\mu^2$, $u_k(x)\in C^{(\infty)}[0,1]$ and $\pi_k(t)\in C^{(\infty)}[0,\mu_o]$, $\mu_o=1/\mu$.

Note that $\pi_k(t)=\pi_k(t,\mu)$ i.e., $\pi_k(t)$ depends also on μ, but this dependence is not indicated.

The initial conditions for the functions $\pi_j(t)$ are taken as,

$$\pi_{-1}(1/\mu)=b\mu,\quad b=u^o-\sum_{k=o}^{\infty}\mu^k u_k(1),\quad \pi_k(\mu_o)=0,\quad k=0,1,\ldots$$

Substituting equation $y(x)=\mu^{-1}\pi_{-1}(t)+\sum_{k=o}^{\infty}\left(\pi_k(t)+u_k(x)\right)\mu^k$, into equation $(x+\varepsilon y(x))y'(x)+q(x)y(x)=r(x)$, $y(1)=b$ we obtain to determine the functions $\pi_k(t), k=-1,0,1,\ldots,\ u_n(x), n=0,1,\ldots,$

we have the following equations:

$$\left(t+\pi_{-1}(t)\right)\pi'_{-1}(t)=q(\mu t)\pi_{-1}(t),\quad \pi_{-1}(\mu_o)=b\mu,$$

$$Lu_o(x):=xu'_o(x)-q(x)u_o(x)=r(x),\quad u_o(x)\in C^{(\infty)}[0,1]$$

$$D\pi_o(t):=\left(t+\pi_{-1}(t)\right)\pi'_o(t)+\left(\pi'_{-1}(t)-q(\mu t)\right)\pi_o(t)=-u_o(t\mu)\pi'_{-1}(t),\quad \pi_o(\mu_o)=0$$

$$Lu_1(x)=0,\quad u_1(x)\in C^{(\infty)}[0,1],$$

$$D\pi_1(t)=-u_o(t\mu)\pi'_o(t)+\pi_o(t)\pi'_o(t)-u_1(t\mu)\pi'_{-1}(t),\quad \pi_1(\mu_o)=0$$

$$Lu_2(x):=-u_o(x)u'_o(x),\quad u_2(x)\in C^{(\infty)}[0,1]$$

$$D\pi_2(t):=-u_o(t\mu)\pi'_{-1}(t)-\pi_o(t)\pi'_1(t)-u_1(t\mu)\pi'_o(t)-\pi_1(t)\pi'_o(t)-u_2(t\mu)\pi'_{-1}$$

$$(t),\pi_2(\mu_o)=0$$

$$Lu_3(x):=-u_o(x)u'_1(x)-u'_o(x)u_1(x),\quad u_3(x)\in C^{(\infty)}[0,1],$$

$$D\pi_3(t)=\sum_{\substack{i+j=2\\ i\geq 0,j\geq -2}} u_i(\mu t)\pi'_j(t)+\sum_{\substack{i+j=2\\ i,j\geq 0}}\pi_i(t)\pi'_j(t),\quad \pi_3(\mu_o)=0,$$

We solve these problems successively. We write equation,

$D\pi_1(t) = -u_0(t\mu)\pi_0'(t) + \pi_0(t)\pi_0'(t) - u_1(t\mu)\pi_{-1}'(t), \quad \pi_1(\mu_0) = 0$ as,

$$tz'(t) - q(\mu t)z(t) = -z(t)z'(t), \; z(\mu_0) = b\mu,$$

Where,

$$z = \pi_{-1}(t), \; \mu_0 = \mu^{-1}.$$

The fundamental solution of the homogeneous equation corresponding to this equation is of the form,

$$z^0(t) = \exp\left\{\int_{\mu_0}^{t} q(\mu s)\frac{ds}{s}\right\} = \exp\left\{\int_{\mu_0}^{t} (q(\mu s)+1)\frac{ds}{s} - \int_{\mu_0}^{t} \frac{ds}{s}\right\} = \frac{p(t,\mu)}{\mu t}$$

where,

$$p(t,\mu) = \exp\left\{\int_{\mu_0}^{t} (q(\mu s)+1)\frac{ds}{s}\right\}.$$

Using the expression for $z^0(t)$, the solution of the inhomogeneous equation for $z(t)$, can be written as,

$$z(t) = \frac{p(t,\mu)}{\mu t}\left[z(\mu_0) + \mu\int_{\mu_0}^{t} p^{-1}(s,\mu)z(s)z'(s)ds\right],$$

$$\text{Or } tz(t) = p(t,\mu)b - p(t,\mu)\int_{\mu_0}^{t} p^{-1}(s,\mu)z(s)z'(s)ds.$$

After integrating by parts, we reduce the last expression to the following equation:

$$tz(t) = p(t,\mu)b - \frac{z^2(t)}{2} + p(t,\mu)\frac{b^2\mu^2}{2} + \frac{p(t,\mu)}{2}\int_{\mu_0}^{t} \frac{1+q(\mu s)}{s}p^{-1}(s,\mu)z^2(s)ds$$

or

$$z^2(t) + 2tz(t) - p(t,\mu)b_0 = p(t,\mu)\int_{\mu_0}^{t} \varphi(s,\mu)p^{-1}(s,\mu)z^2(s)ds := p(t,\mu)T(t,z^2)$$

where $\phi(s,\mu) = (1+q(\mu s))/s, \; b_0 = 2b + b^2\mu^2$

Let $b_0 > 0$. Let us introduce the notation $z_0(t) = -t + \sqrt{t^2 + b_0 p(t,\mu)}$.

This function satisfies the inequality $0 < z_0(t) \le Mt^{-1} \; (t > 0)$ and is a strictly decreasing bounded function on the closed interval $[0, \mu_0]$. Here and elsewhere, all constants independent of the small parameter μ are denoted by M. Let S_μ be the set of functions $z(t)$ satisfying the condition,

$\| z - z_0 \| \le M\mu$, where $\| z \| = \max\limits_{0 \le t \le \mu_0} | z(t) |$,

Theorem: If $o > o$, then there exists a unique constraint of the solution of equation $D\pi_1(t) = -u_o(t\mu)\pi_o'(t) + \pi_o(t)\pi_o'(t) - u_1(t\mu)\pi_{-1}'(t), \quad \pi_1(\mu_o) = o$ from the set S_μ.

Proof: Equation $z^2(t) + 2tz(t) - p(t,\mu)b_o = p(t,\mu)\int_{\mu_o}^{t}\varphi(s,\mu)p^{-1}(s,\mu)z^2(s)ds := p(t,\mu)$ $T(t,z^2)$ is equivalent to the equation $z = F[t,z]$, where,

$$F[t,z] = -t + \sqrt{t^2 + bp(t,\mu) + p(t,\mu)T(t,z^2)}.$$

Suppose that,

$\|\phi(t,\mu)\| \le M\mu$, $o < m \le p(t,\mu) \le M$, $\|p^{-1}(t)\| \le M$. First, let us estimate $T(t,z^2)$ on the set S_μ. We have,

$$\left|T(t,z^2)\right| \le \int_t^{\mu_o}|\phi(s,\mu)||p^{-1}(s,\mu)||z(s)|^2 ds \le M\mu\int_t^{\mu_o}|z(s)|^2 ds \le M\mu\int_o^{\mu_o}|z(s)|^2 ds \le$$
$$\le M\mu\int_o^1|z(s)|^2 ds + M\mu\int_1^{\mu_o}|z(s)|^2 ds \le M\mu.$$

Here, we have used the triangle inequality,

$$|z(t)| \le |z(t) - z_o(t)| + |z_o(t)|,$$

as well as the inequality,

$$|z_o(t)| \le Mt^{-1} \quad (t > o).$$

The Fréchet derivative of the operator $F(t,z)$ with respect to z at the point $z_o(t)$ is a linear operator:

$$F'_z(t,z_o)h = -p(t,\mu)\int_t^{\mu_o}\varphi(s,\mu)p^{-1}(s,\mu)z_o(s)h(s)\frac{ds}{\sqrt{t^2 + p(t,\mu)\left(b + T(t,z^2)\right)}},$$

where $h(t)$ is a continuous function on the closed interval $[o,\mu o]$. Note that, in view of $T(t,z_o^2) = O(\mu)$, the denominator of this expression is strictly positive on the closed interval $[o,\mu_o]$. For $F_z'(t,z_o)$, we can obtain the estimate $\|F_z(t,z_o)\| \le M\mu\ln\mu^{-1}$ in the same way as the estimate for $T(t,z^2)$. Hence, in turn, it follows from the Lagrange inequality that the operator is a contraction operator in the set S_μ. Therefore, by the fixed-point principle, equation $z^2(t) + 2tz(t) - p(t,\mu)b_o = p(t,\mu)\int_{\mu_o}^{t}\varphi(s,\mu)p^{-1}(s,\mu)z^2(s)ds := p(t,\mu)T(t,z^2)$ has a unique solution from the class SS_μ. The theorem is proved.

Corollary: The following inequalities hold,

$z(t) = \pi_{-1}(t) \geq M > 0$ for all $t \in [\,0,\ \mu\ 0\,]$;

$\pi_{-1}(t) \leq Mt^{-1}\ (t > 0)$.

The other function $\pi_j(t),\ u_j(x), j = 0,1,2,\ldots$ is determined from the inhomogeneous linear equations; therefore, the following lemmas are needed.

Lemma: For any function $f(x) \in C(\infty)[0,1]$, the equation $L\,\xi = f(x)$ has a unique bounded solution $\xi(x) \in C(\infty)\ [0,1]$, expressible as,

$$\xi(x) = Q(x)\int_0^x Q^{-1}(s)f(s)\frac{ds}{x}, Q(x) = \exp\left\{\int_1^x (q(s)+1)\frac{ds}{s}\right\}$$

Proof: The proof follows from the fact that the general solution of the equation under consideration is expressed as,

$$\xi(x) = Q(x)x^{-1}\left[\xi(1) + \int_0^x Q^{-1}(s)f(s)ds\right].$$

If we choose,

$$\xi(1) = \int_0^1 Q^{-1}(s)f(s)ds.$$

then we obtain the required result.

This lemma implies that all the functions $u_k(x), k = 0,1,\ldots$ are uniquely determined and belong to the class $C^\infty[0,1]$.

Lemma: The problem,

$$(t + \pi_{-1}(t))\eta'(t) + (\pi'_{-1}(t) - q(\mu t))\eta(t) = k(t), \quad \eta(\mu_0) = 0,$$

where the function $k(t)$ belongs to $C(\infty)[0,1]$ is continuous and bounded, and if $|k(t)| \leq M\,t-2,\ t \to \infty$, has a unique uniformly bounded solution $\eta(t) = \eta(t,\mu)$ on the closed interval $t \in [0,\mu 0]$ for a small μ.

Proof: The fundamental solution of the homogeneous equation $(t + \pi_{-1}(t))\eta'(t) + (\pi'_{-1}(t) - q(\mu t))\eta(t) = k(t), \quad \eta(\mu_0) = 0$, is of the form,

$$\Phi(t) = \frac{(1+\mu^2 b)g(t,\mu)}{\mu(t+\pi_{-1}(t))},\ g(t,\mu) = \exp\left\{\int_t^{\mu_0} (1 + q,(\mu s))\frac{ds}{s + \pi_{-1}(s)}\right\}.$$

Obviously, $\|g(t,\mu)\| \le M$ and $g^{-1}(t,\mu) \le M$ for $t \in [0,\mu_0]$ and μ are small. The solution of equation $(t+\pi_{-1}(t))\eta'(t)+(\pi'_{-1}(t)-q(\mu t))\eta(t)=k(t),\quad \eta(\mu_0)=0$, can be expressed as,

$$\eta(t)=\frac{g(t,\mu)}{t+\pi_{-1}(t)}\int_{\mu_0}^{t} g^{-1}(s,\mu)k(s)ds.$$

The estimate of the integral term in equation above shows that it is bounded by the constant M. Hence, it also follows that $|\eta(t)| \le Mt^{-1}\,(t>0)$. The solution of equation $D\pi_0(t):=(t+\pi_{-1}(t))\pi'_0(t)+(\pi'_{-1}(t)-q(\mu t))\pi_0(t)=-u_0(t\mu)\pi'_{-1}(t),\quad \pi_0(\mu_0)=0$ is defined by the integral equation above where,

$$k(t)=-u_0(t\mu)\pi_{-1}(t)=-u_0(t\mu)q(\mu t)\pi_{-1}(t)t+\pi_{-1}(t),$$

satisfies the assumptions of the lemma. Therefore, the function $\pi_0(t)$ is bounded on $[0,\mu_0]$. The boundedness of the other functions $\pi_k(t), k=1,2,\ldots$ is proved in a similar way, because the right-hand sides of the equations defining these functions satisfy the assumptions of this Lemma. The estimate of the asymptotic behavior of the equation $y(x)=\mu^{-1}\pi_{-1}(t)+\sum_{k=0}^{\infty}(\pi_k(t)+u_k(x))\mu^k$, is also carried out using this Lemma.

Let us introduce the notation,

$$y(x)=\mu^{-1}\pi_{-1}(t)+\sum_{k=0}^{n}\mu^k(\pi_k(t)+u_k(x))+\mu^{n+1}R_{n+1}(x,\mu).$$

The following statement holds.

Theorem: Let $bo>0$ (for this, it suffices that the condition $bo:=b-yo(1)>0$ holds). Then the solution of problem (76) exists on the closed interval [0,1] and its asymptotics can be expressed as equation above and $|Rn+1(x,\mu)| \le M$ for all $x \in [0,1]$.

Example: Consider the equation,

$$(x+\varepsilon y(x))y'(x)+y(x)=1,\quad y(1)=b,$$

This equation is integrated exactly,

$$y(x)=\varepsilon^{-1}\left[-x+\sqrt{x^2+2b_0\varepsilon+\varepsilon^2\left(y^{(0)}\right)^2+2\varepsilon x}\right],$$

Where $b_0=b-1$. If $b_0>0$, then the solution of equation $y'(x)=f(x,y,\varepsilon),\ y(0)=0$ exists on the closed interval [0, 1], which is confirmed by Theorem. The equation for $\pi_{-1}(t)$ is of the form,

$$(t+\pi_{-1}(t))\pi'_{-1}(t)+\pi_{-1}(t)=0,\ \pi_{-1}(\mu_0)=b\mu.$$

The solution of this problem can be expressed as,

$$\pi_{-1}(t)=-t+\sqrt{t^2+2b+b^2\mu^2}.$$

The equation for $u0(x)$ has the solution $y_0(x)=1\in C^\infty[0,1]$. Further,

$$\pi_0(t)=\frac{-\pi_{-1}(t)+b\mu}{t+\pi_{-1}(t)},\ u_k(x)=0,\ k=1,2,\ldots,$$

where $b=b_0$. The asymptotics of the solutions of equation $(x+\varepsilon y(x))y'(x)+q(x)y(x)=r(x),\ y(1)=b$ can be expressed as,

$$y(x)=\mu^{-1}\pi^{-1}(x/\mu)+1+\pi_0(x/\mu)+o(\mu)\ \text{for all}\ x\in[0,1],\ \mu\to 0.$$

Lagerstrom Model Problem

The problem,

$$v''(r)+\frac{k}{r}v'(r)+v(r)v'(r)=\beta[v'(r)]^2,\ v(\varepsilon)=0,\ v(\infty)=1,$$

where $0<\beta$ is constant, $k\in N$.

It has been proposed as a model for Lagerstrom Navier-Stokes equations at low Reynolds numbers. It can be interpreted as a problem of distribution of a stationary temperature v(r).

The first two terms in equation $y'(x)=f(x,y,\varepsilon),\ y(0)=0$ is (k+1) dimensional Laplacian depending only on the radius, and the other two members—some nonlinear heat loss.

It turns out that not only the asymptotic solution but also convergent solutions of equation $y'(x)=f(x,y,\varepsilon),\ y(0)=0$ can be easily constructed by a fictitious parameter. The basic idea of this method is as follows. The initial problem is entered fictitious parameter $\lambda\in[0,1]$ with the following properties:

$\lambda=0$, the solution of the equation satisfies all initial and boundary conditions;

The solution of the problem can be expanded in integral powers of the parameter λ for all $\lambda\in[0,1]$.

It is convenient in equation $v''(r)+\frac{k}{r}v'(r)+v(r)v'(r)=\beta[v'(r)]^2,\ v(\varepsilon)=0,\ v(\infty)=1,$ to make setting $r=\varepsilon x, v=1-u$, then,

$$u''(x)+\left(kx^{-1}+\varepsilon\right)u'(x)-\lambda\varepsilon u(x)u'(x)=[u'(x)]^2,\ u(1)=1,\ u(\infty)=0.$$

We have the following:

Theorem: For small $\varepsilon>0$, the solution of equation above can be represented in the form of absolutely and uniformly convergent series,

$$u(x)=u_0(x,\varepsilon)+v_k(\varepsilon)u_1(x,\varepsilon)+\ldots+v_k^n(\varepsilon)u_n(x,\varepsilon)+\ldots,$$

for the sufficiently small parameter ε, where,

$$v_1(\varepsilon)\sim\left(\ln\frac{1}{\varepsilon}\right)-1,\ v_2\sim\varepsilon\ln\frac{1}{\varepsilon},\ v_k\sim\frac{k-1}{k-2}\varepsilon(j>2)\ ;u_k(x,\varepsilon)=O(1),\forall x\in[1,\infty)$$

Note that the function $u_n(x,\varepsilon)$ also depends on k, but for simplicity, this dependence is not specified.

Proof: We introduce equation $u''(x)+\left(kx^{-1}+\varepsilon\right)u'(x)-\lambda\varepsilon u(x)u'(x)=[u'(x)]^2$, $u(1)=1$, $u(\infty)=0$. parameter λ, i.e., consider the problem,

$$u''(x)+\left(kx^{-1}+\varepsilon\right)u'(x)-\beta[u'(x)]^2=\lambda\varepsilon u(x)u'(x),\ u(1)=1,\ u(\infty)=0$$

Setting $\lambda = 0$ in equation above we have,

$$u_0''+\left(x^{-1}k+\varepsilon\right)u_0'=0,\ u_0(1)=1,\ u_0(\infty)=0.$$

It has a unique solution,

$$u_0=X(x,\varepsilon):=1-X_1(X,\varepsilon),\ X_1=C_0\int_1^x s^{-k}e^{-\varepsilon s}ds,\ C_0^{-1}=\int_1^\infty s^{-k}e^{-\varepsilon s}ds.$$

Therefore, equation $u_0''+\left(x^{-1}k+\varepsilon\right)u_0'=0,\ u_0(1)=1,\ u_0(\infty)=0$ with zero boundary conditions is the Green's function,

$$K(x,s,\varepsilon)=\begin{cases}C_0^{-1}X_1(x,\varepsilon)X(s,\varepsilon), & 1\le x\le s,\\ C_0^{-1}X_1(s,\varepsilon)X(x,\varepsilon), & s<x<\infty.\end{cases}$$

Hence, the equation $u''(x)+\left(kx^{-1}+\varepsilon\right)u'(x)-\beta[u'(x)]^2=\lambda\varepsilon u(x)u'(x),\ u(1)=1,\ u(\infty)=0$ is reduced to the system of integral equations,

$$u(x)=X(x,\varepsilon)+\lambda\varepsilon\int_1^\infty G(x,s,\varepsilon)u(s)u'(s)ds,$$

$$u'(x)=X'(x,\varepsilon)+\lambda\varepsilon\int_1^\infty G_x(x,s,\varepsilon)u(s)u'(s)ds,$$

where,

$$G(x,s,\varepsilon)=\begin{cases} X_1(x,\varepsilon)X(s,\varepsilon)/X'(s,\varepsilon), & 1\le x\le s,\\ X_1(s,\varepsilon)X(x,\varepsilon)/X'(s,\varepsilon), & s<x<\infty. \end{cases}$$

In equation,

$$u(x)=X(x,\varepsilon)+\lambda\varepsilon\int_1^\infty G(x,s,\varepsilon)u(s)u'(s)ds,$$
$$u'(x)=X'(x,\varepsilon)+\lambda\varepsilon\int_1^\infty G_x(x,s,\varepsilon)u(s)u'(s)ds,$$

we make the substitution $u=X(x,\varepsilon)\phi(x)$, $u'=X'(x,\varepsilon)\psi(x)$, and then we have,

$$\phi(x)=1+\lambda\varepsilon\int_1^\infty Q_1(x,s,\varepsilon)\phi(s)\psi(s)ds:=1+\lambda\varepsilon Q_1(\phi\psi),$$

$$\psi(x)=1+\lambda\varepsilon\int_1^\infty Q_2(x,s,\varepsilon)\phi(s)\psi(s)ds:=1+\lambda\varepsilon Q_2(\lambda\psi),$$

where,

$$Q_1=X^{-1}(x,\varepsilon)G(x,s,\varepsilon)X(s,\varepsilon)X'(s,\varepsilon),$$
$$Q_2=X_X^{-1}(x,\varepsilon)G_x(x,s,\varepsilon)X(s,\varepsilon)X'(s,\varepsilon).$$

Lemma: The following estimate holds,

$$\int_1^\infty |Q_j(x,s,\varepsilon)|ds\le\int_1^\infty X(s,\varepsilon)ds \quad (j=1,2)$$

Given that, we have,

$0\le X_1(x,\varepsilon)\le 1,\ |X'(x,\varepsilon)|=X'(x,\varepsilon),\ X'(x,\varepsilon)\le 0,\ x\in[1,\infty)$, we have,

$$\int_1^\infty |Q_1(x,s,\varepsilon)|ds\le\int_1^x \frac{X_1(s,\varepsilon)}{X'_1(s,\varepsilon)}|X'(s,\varepsilon)|X(s,\varepsilon)ds+$$

$$\int_x^\infty X^{-1}(x,\varepsilon)\frac{X^2(s,\varepsilon)|X'(s,\varepsilon)|}{X'_1(s,\varepsilon)}ds\le\int_1^x X(s,\varepsilon)ds+\int_x^\infty X(s,\varepsilon)ds=\int_1^\infty X(s,\varepsilon)ds.$$

Inequality equation $\int_1^\infty |Q_j(x,s,\varepsilon)|ds\le\int_1^\infty X(s,\varepsilon)ds \quad (j=1,2)$ for j = 2 is proved similarly.

Further, by integrating by parts, we have,

$$\int_1^\infty X(s,\varepsilon)ds=-1+C_o\int_1^\infty s^{-k+1}e^{-\varepsilon s}ds\le\int_1^\infty s^{-k+1}e^{-\varepsilon s}ds/\int_1^\infty s^{-k}e^{-\varepsilon s}ds:=\frac{v_k(\varepsilon)}{\varepsilon}.$$

Consequently,

$$\varepsilon\int_1^\infty X(x,\varepsilon)ds \le v_k(\varepsilon).$$

It is from integral expressing of $v_k(\varepsilon)$ we can obtain the asymptotic behavior such as indicated in the theorem.

With the solution of equation $\phi(x)=1+\lambda\varepsilon\int_1^\infty Q_1(x,s,\varepsilon)...ds:=1+\lambda\varepsilon Q_2(\lambda\psi),$, we can expand in series,

$$\phi(x)=1+\phi_1(x,\varepsilon)\lambda+\phi_2(x,\varepsilon)\lambda^2+\ldots,$$
$$\psi(x)=1+\psi_1(x,\varepsilon)\lambda+\Psi_2(x,\varepsilon)\lambda^2+\ldots.,$$

The coefficients of this series are uniquely determined from the equations,

$$\phi_o=\Psi_o=1,\ \ \phi_1=\varepsilon Q_1(1),\ \ \Psi_1=Q_2(1),$$
$$\phi_n=\varepsilon Q_1(\phi_{n-1})+\varepsilon Q_1(\Psi_{n-1})+\varepsilon Q_1(\phi 1\Psi_{n-2})+\ldots+\varepsilon Q_1(\phi_{n-2}\Psi_1),$$
$$\Psi_n=\varepsilon Q_2(\phi_{n-1})+\varepsilon Q_2(\Psi_{n-1})+\varepsilon Q_2(\phi_1\Psi_{n-2})+\ldots+\varepsilon Q_2(\phi_{n-2}\Psi_1),\ (n=2,3,\ldots).$$

Let $z=\sup_{1\le x\infty}\{|\phi(x)|,|\Psi(x)|\}$, then by using equation $\varepsilon\int_1^\infty X(x,\varepsilon)ds\le v_k(\varepsilon)$ we have a Majorant equation: $z=1+\lambda_{vk}(\varepsilon)z^2$. The solution of this equation can be expanded in powers λ the under condition $8v_k(\varepsilon)\le 1$ for all $\lambda\in[0,1]$.

If we call $u_n(x,\varepsilon)=\dfrac{X(x,\varepsilon)\phi_n(x,\varepsilon)}{v_k^n(\varepsilon)}$, we get the proof of the theorem.

References

- Differential-equation, maths: byjus.com, Retrieved 28 March, 2020
- 1-solving-des, differential-equations: intmath.com, Retrieved 13 August, 2020
- A-review-on-fractional-differential-equations-and-a-numerical-method-to-solve-some-boundary-value-pr: intechopen.com, Retrieved 18 July, 2020
- Derivative-mathematics, science: britannica.com, Retrieved 27 April, 2020
- Perturbed-differential-equations-with-singular-points, recent-studies-in-perturbation-theory: intechopen.com, Retrieved 16 May, 2020

2

Ordinary Differential Equations

The type of differential equation that contains one or more functions of an independent variable is termed as ordinary differential equation. First-order ordinary differential equation, second-order ordinary differential equation, Abel's identity, Bernoulli differential equation, etc. are studied under its domain. The topics elaborated in this chapter will help in gaining a better perspective about ordinary differential equations.

An ordinary differential equation (frequently called an "ODE," "diff eq," or "diffy Q") is an equality involving a function and its derivatives. An ODE of order n is an equation of the form,

$$F(x,y,y',...,y^{(n)})=0,$$

where y is a function of x, $y'=dy/dx$ is the first derivative with respect to x, and $y^{(n)}=d^n y/dx^n$ is the nth derivative with respect to x.

An ODE of order n is said to be linear if it is of the form,

$$a_n(x)y^{(n)}+a_{(n-1)}(x)y^{(n-1)}+...+a_1(x)y'+a_0(x)y=Q(x).$$

A linear ODE where $Q(x)=0$ is said to be homogeneous. Confusingly, an ODE of the form,

$$y'=f\left(\frac{y}{x}\right)$$

is also sometimes called "homogeneous."

In general, an nth-order ODE has n linearly independent solutions. Furthermore, any linear combination of linearly independent functions solutions is also a solution.

Simple theories exist for first-order (integrating factor) and second-order (Sturm-Liouville theory) ordinary differential equations, and arbitrary ODEs with linear constant coefficients can be solved when they are of certain factorable forms. Integral transforms such as the Laplace transform can also be used to solve classes of linear ODEs. Morse and Feshbach give canonical forms and solutions for second-order ordinary differential equations.

While there are many general techniques for analytically solving classes of ODEs, the only practical solution technique for complicated equations is to use numerical methods. The most popular of these is the Runge-Kutta method, but many others have been developed, including the collocation method and Galerkin method. A vast amount of research and huge numbers of publications have been devoted to the numerical solution of differential equations, both ordinary and partial (PDEs) as a result of their importance in fields as diverse as physics, engineering, economics, and electronics.

The solutions to an ODE satisfy existence and uniqueness properties. These can be formally established by Picard's existence theorem for certain classes of ODEs. Let a system of first-order ODE be given by:

$$\frac{dx_i}{dt} = f_i(x_1,\ldots,x_n,t),$$

for i = 1, ...,n and let the functions $f_i(x_1,\ldots,x_n,t)$, where i = 1, ...,n all be defined in a domain D of the $(n+1)$-dimensional space of the variables $x_1,\ldots,x_n,t$. Let these functions be continuous in D and have continuous first partial derivatives $\partial f_i / \partial x_j$ for i = 1, ...,n and j = 1, ...,n in D. Let $(x_1^o,\ldots,x_n^o)$ be in D. Then there exists a solution of $\frac{dx_i}{dt} = f_i(x_1,\ldots,x_n,t)$, given by:

$$x_1 = x_1(t),\ldots,x_n = x_n(t)$$

For $t_o - \delta < t < t_o + \delta$ (where $\delta > 0$) satisfying the initial conditions,

$$x_1(t_o) = x_1^o,\ldots,x_n(t_o) = x_n^o.$$

Furthermore, the solution is unique, so that if,

$$x_1 = x_1^*(t),\ldots,x_n = x_n^*(t)$$

Other special first-order types include cross multiple equations,

$$y f(xy)dx + x g(xy)dy = 0,$$

homogeneous equations,

$$y' = f\left(\frac{y}{x}\right),$$

linear equations,

$$y' + p(x)y = q(x),$$

and separable equations,

$$y' = X(x)Y(y).$$

Special classes of second-order ordinary differential equations include:

$$y'' = f(y, y')$$

(x missing) and

$$y'' = f(x, y')$$

(y missing). A second-order linear homogeneous ODE,

$$y'' + P(x)y' + Q(x)y = 0$$

for which,

$$\frac{Q'(x) + 2P(x)Q(x)}{2[Q(x)]^{3/2}} = [\text{constant}]$$

can be transformed to one with constant coefficients.

The undamped equation of simple harmonic motion is:

$$y'' + \omega_0^2 y = 0,$$

which becomes,

$$y'' + \beta y' + \omega_0^2 y = 0$$

when damped, and

$$y'' + \beta y' + \omega_0^2 y = A\cos(\omega t + \delta)$$

when both forced and damped.

Systems with constant coefficients are of the form,

$$\frac{dx}{dt} = A\mathbf{x}(t) + \mathbf{p}(t).$$

The following are examples of important ordinary differential equations which commonly arise in problems of mathematical physics.

Abel's differential equation:

$$y = f_0(x) + f_1(x)y + f_2(x)y^2 + f_3(x)y^3 + \dots$$

$$[g_0(x) + g_1(x)y]y' = f_0(x) + f_1(x)y + f_2(x)y^2 + f_3(x)y^3.$$

Airy differential equation:

$$y'' - xy = 0.$$

Anger differential equation:

$$y'' + \frac{y'}{x} + \left(1 - \frac{v^2}{x^2}\right)y = \frac{x - v}{\pi x^2}\sin(v\pi).$$

Baer differential equations:

$$(x-a_1)(x-a_2)y''+\frac{1}{2}[2x-(a_1+a_2)]y'-(p^2x+q^2)y=0,$$

$$(x-a_1)(x-a_2)y''+\frac{1}{2}[2x-(a_1+a_2)]y'-(k^2x^2-p^2x+q^2)y=0.$$

Bernoulli differential equation:

$$y'+p(x)y=q(x)y^n.$$

Bessel differential equation:

$$x^2y''+xy'+(\lambda^2x^2-n^2)y=0.$$

Binomial differential equation:

$$(y')^m=f(x,y).$$

Bôcher equation:

$$y''+\frac{1}{2}\left[\frac{m_1}{x-a_1}+\ldots+\frac{m_{n-1}}{x-a_{n-1}}\right]y'+\frac{1}{4}\left[\frac{A_0+A_1x+\ldots+A_lx^l}{(x-a_1)^{m_1}(x-a_2)^{m_2}\ldots(x-a_{n-1})^{m_{n-1}}}\right]$$

Briot-Bouquet equation:

$$x^my'=f(x,y).$$

Chebyshev differential equation:

$$(1-x^2)y''-xy'+\alpha^2y=0.$$

Clairaut's differential equation:

$$y=xy'+f(y').$$

Confluent hypergeometric differential equation:

$$xy''+(c-x)y'-ay=0.$$

d'Alembert's equation:

$$y=xf(y')+g(y').$$

Duffing differential equation:

$$y''+\omega_0^2y+\beta y^3=0.$$

Eckart differential equation:

$$y''+\left[\frac{\alpha\eta}{1+\eta}+\frac{\beta}{(1+\eta)^2}+\gamma\right]y=0,$$

Where,

$$\eta = e^{(\delta x)}.$$

Emden-Fowler differential equation:

$$(x^p y')' + / - x^\sigma y^n = 0.$$

Euler differential equation:

$$x^2 y'' + a x y' + by = S(x).$$

Halm's differential equation:

$$(1+x^2)^2 + y'' + "\lambda y = 0.$$

Hermite differential equation:

$$y'' - 2x y' + \lambda y = 0.$$

Heun's differential equation:

$$w'' + \left(\frac{\gamma}{x} + \frac{\delta}{x-1} + \frac{\in}{x-a}\right) w' + \frac{\alpha\beta x - q}{x(x-1)(x-a)} w = 0$$

Where $w' = dw / dx$.

Hill's differential equation:

$$y'' + \left[\theta_0 + 2\sum_{n=1}^{\infty} \theta_n \cos(2nz)\right] y = 0$$

Hypergeometric differential equation:

$$x(x-1)y'' + [(1+\alpha+\beta)x - \gamma]y' + \alpha\beta y = 0.$$

Jacobi differential equation:

$$(1-x^2)y'' + [\beta - \alpha - (\alpha+\beta+2)x]y' + n(n+\alpha+\beta+1)y = 0.$$

Laguerre differential equation:

$$x y'' + (1-x)y' + \lambda y = 0.$$

Lamé's differential equation:

$$(x^2 - b^2)(x^2 - c^2)z'' + x(x^2 - b^2 + x^2 - c^2)z' - [m(m+1)x^2 - (b^2 + c^2)p]z = 0,$$

where $z' = dz / dx$.

Lane-Emden differential equation:

$$\frac{1}{\xi^2}\frac{d}{d\xi}\left(\xi^2\frac{d\theta}{d\xi}\right)+\theta^n=0.$$

Legendre differential equation:

$$(1-x^2)y''-2xy'+\alpha(\alpha+1)y=0.$$

Linear constant coefficients:

$$a_0y^{(n)}+\ldots+a_{(n-1)}y'+a_ny=p(x).$$

Lommel differential equation:

$$x^2y''+xy'-(x^2+v^2)y=k\,x^{(\mu+1)}.$$

Löwner's differential equation:

$$y'=-y\frac{1+\kappa(x)y}{1-\kappa(x)y}$$

Malmstén's differential equation:

$$y''+\frac{r}{z}y'=\left(Az^m+\frac{s}{z^2}\right)y.$$

Mathieu differential equation:

$$V''+[a-2q\cos(2v)]V=0,$$

where $V'=dV/dv$.

Modified Bessel differential equation:

$$x^2y''+xy'-(x^2+n^2)y=0.$$

Modified spherical Bessel differential equation:

$$r^2R''+2rR'-[k^2r^2+n(n+1)]R=0,$$

where $R'=dR/dr$

Rayleigh differential equation:

$$y''-\mu\left(1-\frac{1}{3}y'^2\right)y'+y=0.$$

Riccati differential equation:

$$w'=q_0(x)+q_1(x)w+q_2(x)w^2.$$

Riemann P-Differential Equation:

$$u'' + \left[\frac{1-\alpha-\alpha'}{z-a} + \frac{1-\beta-\beta'}{z-b} + \frac{1-\gamma-\gamma'}{z-c}\right]u'$$

$$+\left[\frac{\alpha\alpha'(a-b)(a-b)(a-c)}{z-a} + \frac{\beta\beta'(b-c)(b-a)}{z-b} + \frac{\gamma\gamma'(c-a)(c-b)}{z-c}\right]$$

$$\frac{u}{(z-a)(z-b)(z-c)} = 0$$

Where $u' = du/dz$.

Sharpe's differential equation:

$$zy'' + y' + (z + A)y = 0.$$

Spherical Bessel differential equation:

$$r^2R'' + 2rR'' + [k^2r^2 - n(n+1)]R = 0,$$

where $R' = dR/dr$.

Struve differential equation:

$$z^2y'' + zy' + \left(z^2 - v^2\right)y = \frac{a\left(\frac{1}{2}z\right)^{v+1}}{\sqrt{\pi}\Gamma\left(v + \frac{1}{2}\right)}.$$

Sturm-Liouville equation:

$$\frac{d}{(dx)}[p(x)y'] + [\lambda w(x) - q(x)]y = 0.$$

Gegenbauer differential equation:

$$(1-x^2)y'' - (2\alpha+1)xy' + n(n+2\alpha)y = 0.$$

van der Pol equation:

$$y'' - \mu(1-y^2)y' + y = 0.$$

Weber differential equation:

$$y'' + \left(n + \frac{1}{2} - \frac{1}{4}z^2\right)y = 0,$$

Where $y'' = dy/dz$.

Whittaker differential equation:

$$u'' + u' + \left(\frac{k}{z} + \frac{\frac{1}{4} - m^2}{z^2} \right) u = 0$$

Where $u' = du/dz$.

SEPARABLE EQUATIONS

A differential equation is said to be separable if the variables can be separated the theorem behind integration by substitution from calculus.

Which states that if $f(x)$ is a continuous function and y has a derivative that is a continuous function - hence y being a function itself, then integrating such chained functions yields:

$$\int f(y(x))y'dx = \int f(y)dy$$

In other words, our integrand $f(y(x))y'(x)$ and our antiderivate will be the same if we took the antiderivative of f and plugged it into $y(x)$. This provides a very useful tool to solve many ODE's.

To take an example let's consider the differential equation:

$$y' = ay$$

We comment the constant function $y(x) = 0$ is a solution. We will focus on finding solutions that are not zero. As commented before it is a consequence of the uniqueness of solutions that two solutions can never intersect. For this problem it means that solutions are always positive or always negative because they cannot cross $y(x) = 0$. For easier understanding, picture a normal Cartesian coordinate system. The solutions can only be above or below the x-axis.

For this reason we are justified for the remainder in assuming that $y(x) \neq 0$. Thus we can divide by $y(x)$ as division by zero is undefined, to find that:

$$\frac{1}{y}y' = a$$

Now integrating both sides with respect to x we have that:

$$\int \frac{1}{y}y'dx = \int a dx$$

Using the theorem about integration by substitution theorem:

$$\int \frac{1}{y} dy = \int a dx$$

Or

$$\ln|y| = ax + C.$$

$$y = \pm e^C e^{ax} = Ce^{ax}.$$

Finally we comment that, $e^c > 0$ and e^c collapsing to C, so technically speaking the above derivation only shows that we get a solution if $C > 0$ or if $C < 0$. Taking into account $\pm$.

Notice on the other hand that if we allow $C = 0$ then we can recover the solution $y(x) = 0$. Thus we can represent all solutions as,

$$y(x) = Ce^{ax}$$

where C is a real number.

This method is known as separation of variables because after we have divided we have one side that is entirely in terms of y and the other side entirely in terms of x.

Notation and Heuristics

While the change of variables theorem is the mathematical justification, using Leibniz' notation for derivatives allows for excellent mnemonics for remembering how separation of variables works.

We present two notationally different but equivalent derivations here:

Beginning from the point where we have divided by $y(x)$ and using Leibniz notation we have that:

$$\frac{1}{y}\frac{dy}{dx} = a.$$

One might not view this as fully separated because there is still a "dx" on the left hand side. But one can fully separate the equation by multiplying by "dx", to get:

$$\frac{1}{y} dy = a dx$$

Then placing an integral in front of each term one gets:

$$\int \frac{1}{y} dy = \int a dx.$$

This arrives at the correct place, but treating "dx" as a number is purely heuristic.

A second mnemonic which is fairly similar uses Leibniz notation to remind us how the change of variables theorem should work. Beginning from the step in the original example where we integrated both sides with respect to x, and using Leibniz notation we have:

$$\int \frac{1}{y}\frac{dy}{dx}dx = \int a dx.$$

Now we can "cancel the dx" to get that,

$$\int \frac{1}{y}\frac{dy}{\cancel{dx}}\cancel{dx} = \int \frac{1}{y}dy.$$

Thus,

$$\int \frac{1}{y}dy = \int a dx.$$

Again, "dx" is not a number, so it cannot really be canceled. But the heuristic does correctly give the change of variables formula.

An ordinary differential equation is called separable if it can be rewritten in the form:

$$M(x) + N(y)y' = 0$$

Integrating both sides with respect to x:

$$\int M(x)dx + \int N(y)y'dx = C$$

Now by one of the methods we can change and simplfy the second integral to,

$$\int M(x)dx + \int N(y)dy = C$$

An equation where you can factor out both P and Q into separate functions of x and y,

$$P_1(x)P_2(y)dx + Q_1(x)Q_2(y)dy = 0$$

is called separable because the equation can become an equation with separated variables.

$$\frac{P_1(x)}{Q_1(x)}dx + \frac{Q_2(y)}{P_2(y)}dy = 0$$

$$\frac{P_1(x)}{Q_1(x)}dx = -\frac{Q_2(y)}{P_2(y)}dy$$

However, attention must be paid to this process of division. The cases when $Q_1(x) = 0$ and $P_2(y) = 0$ will be additional possible integral solutions that will not be found through the integral because of this division.

There are two special cases of a separable equation that make the solution almost trivial. These are the cases where either,

$$N(y) = k$$

or

$$M(x) = k,$$

where k is a constant. In either case, the equation changes from being in terms of 3 variables and reduces down to 2. When we can throw out one of the two variables, the solution can be achieved by a simple integral.

No *y* Term: *N*(*y*) = *k*

If $N(y)=k$, we can treat k as part of $M(x)$, and turn $N(y)$ into 1. That is, if we divide both sides by k and let $f(x) = M(x)/k$. Then we're left with,

$$\frac{dy}{dx} = M(x)/k = f(x).$$

This is now basic calculus - take the integral of both sides. The general solution is,

$$y = \int f(x)dx + C,$$

where C is a constant.

Example:

$$\frac{dy}{dx} = x^2 + 7$$

Here, we have a derivative equaling a function of x. The original function is the antiderivative of this, that is, we have to integrate to find it. Now we integrate both sides of the equation, w.r.t. x (with regard to 'x'):

$$\int \frac{dy}{dx} dx = \int (x^2 + 7)dx$$

$$y = \frac{1}{3}x^3 + 7x + C$$

This is the general solution. A function that solves a differential equation is said to satisfy it, so,

$$y = \frac{1}{3}x^3 + 7x + C$$

satisfies the differential equation:

$$\frac{dy}{dx} = x^2 + 7.$$

Example:

$$\frac{dy}{dx}=\sin x+e^{x}$$

This works the same as before - integrate both sides with respect to *x*.

$$\int\frac{dy}{dx}dx=\int\sin x+e^{x}dx$$

$$y=-\cos x+e^{x}+C$$

Note that even in this simple case, it may be impossible to write down a formula for the solution because *f*(*x*) cannot be integrated in elementary terms. In this case, writing the integral of *y*, which is the function we are looking for, symbolically with an integral symbol is deemed to be a solution.

No *x* Term: *M*(*x*)=*k*

Another special case is when,

$$M(x)=k.$$

The only difference is how we make it into an integrable form. Our equation is,

$$N(y)\frac{dy}{dx}=-k.$$

This type of equation is called an autonomous differential equation. The differential equation has no explicit dependence on the independent variable *x* except through the function *y*.

Integrating both sides with respect to *x* we get,

$$\int N(y)\frac{dy}{dx}dx=\int kdx.$$

We now have two integrals, each with respect to the functions contained in them. The solution is of the form,

$$\int N(y)dy=kx+C.$$

You can now try to solve for *y*. It is possible that the left hand side doesn't have an antiderivative that can be expressed as a nice formula, and in that case, we leave it in a as clean a form as possible, that is with the integral symbol unresolved. Another important concept emerges here.

Example:

$$\frac{dy}{dx}=\frac{k}{y}$$

Here we multiply through by y, and multiply through by dx. The rest of the math works as before.

$$ydy = kdx$$

$$\int ydy = \int kdx$$

$$\frac{1}{2}y^2 = kx + C$$

$$y^2 = 2kx + C.$$

In this case, its more convenient to solve for $y2$ than y. If we take the square root to solve for y, we change the equation (since the square root of a number is always positive, we would lose half of each solution curve). In these cases its just fine to leave it in its current form.

You may have noticed that the right hand side of the equation, before solving for y, is the same for both examples. When $f(x)=k$, the right hand side will always be $kx+C$ where k is whatever constant is in the problem. It only changes when $f(x)$ is not constant.

As before, it may be impossible to do the integral. Even if it is possible, it may then not be possible to invert the equation to give y explicitly as a function of x, for example if we get tan(y) + 1/y = x. Again, this does not matter; the result is regarded as a valid solution.

General Separable Variable Equations

In the most general case, neither $M(x)$ or $N(y)$ is a constant. In this case, we use the same method as before (although the variables might have to be separated first). The only difference is that both sides result in non-trivial integrals, so we have a little more work to do. This can produce some untidy solutions, and ODEs are known for this. This untidiness often means that we might want to check our result at the end.

Example:

$$\frac{dy}{dx} = y^3 x$$

Dividing by y^3 and multiplying by dx gives,

$$\frac{dy}{y^3} = xdx$$

$$\int \frac{dy}{y^3} = \int xdx$$

$$-\frac{1}{2y^2} = \frac{1}{2}x^2 + C$$

$$y^2 = -\frac{1}{x^2 + C}$$

Example:

$$\frac{dy}{dx} = \frac{y}{x}$$

We multiply by $\frac{dx}{y}$:

$$\frac{dy}{y} = \frac{dx}{x}$$

$$\int \frac{dy}{y} = \int \frac{dx}{x}$$

$$\ln(y) = \ln(x) + C$$

$$y = e^{\ln(x)+C} = e^{C}e^{\ln(x)} = Cx$$

Again e^c collapsing into C.

Example:

$$\frac{dy}{dx} = xy + y$$

This doesn't look separable at first glance. A little factoring can be applied though:

$$\frac{dy}{dx} = (x+1)y$$

This is separable,

$$\frac{dy}{y} = (x+1)dx$$

$$\int \frac{dy}{y} = \int (x+1)dx$$

$$\ln(y) = \frac{1}{2}x^2 + x + C$$

Problems with Initial Conditions

Solving an initial value problem is fairly easy. Solve for y as we did above. Then once you have the equation, plug in b for y and a for x.

Then, solve for the constant.

Example:

$$\frac{dy}{dx} = ky, \quad y(0) = 10$$

As we saw in previous example, the general solution to this is $y=Ce^{kx}$. Let's plug in our boundary conditions:

$$10 = Ce^{0k}$$

$$10 = Ce^{0}$$

$$C = 10$$

$$y = 10e^{kx}$$

This is the particular solution.

Unfortunately in some cases care must be used when solving for the constant.

Example:

$$2yy' = 2x + 2 \qquad y(0) = -2$$

This equation is already separated, so integrating both sides,

$$\int 2yy'dx = \int 2x + 2dx$$

$$y^2 = x^2 + 2x + C$$

By plugging in we find that:

$$C = (-2)^2 = 4$$

So now we have that:

$$y^2 = x^2 + 2x + 4.$$

But if we continue to solve for y we must be careful. One may naively write: $y = \pm\sqrt{x^2 + 2x + 4}$.

But there is a problem, we are looking for a *function* $y(x)$, which means the function must be single valued. And we should be careful what we mean by ±. Do we mean that for each different x there is a different possibility of a sign?

Fortunately since the function has to be differentiable (and hence continuous), the function $y(x)$ cannot jump back and forth between positive and negative values. So we must mean there is one choice of sign:

$$y = \sqrt{x^2 + 2x + 4} \text{ ; Or}$$

$$y = -\sqrt{x^2 + 2x + 4}.$$

It is a nice consequence of uniqueness, that only one of the two functions could solve the problem. Both solve the differential equation, so we can double check the initial condition. In the first instance $y(0)=2$, so that is not our solution, but for the second function $y(0)=-2$ as desired.

Thus the solution is given by:

$$y = -\sqrt{x^2 + 2x + 4}.$$

Example:

Another interesting example that illustrates problems that may arise is:

$$y' = 6xy^{2/3} \qquad y(0) = 0.$$

In order to solve this initial value problem we begin by separating variables.

$$\int y^{-2/3} y' dx = \int 6x dx$$
$$3y^{1/3} = 3x^2 + C$$
$$y = (x^2 + C)^3$$

Now to plug in the initial conditions we have that:

$$0 = (0^2 + C)^3, \text{ Thus}$$
$$C = 0$$

So we have a solution $y(x) = x^6$. But by inspection we also have the solution $y(x) = 0$. So there are at least two solutions to this problem. Perhaps there are more, at the moment we have no way to be sure. Worse, there is nothing that comes up while separating variables to indicate we are in a situation where the situations are not unique.

For example, often one believes a real life process is described by a function which satisfies an ODE. Usually using the ODE, and actual measurements of the process you'd like to deduce what function is governing the process so you can make predictions about future behaviour. But if the solutions to the ODE are not unique, it is difficult to tell when you have found the one solution which is really giving the behaviour of the real world process.

Example:

In the example above we saw that sometimes a differential equation is not well behaved and admits multiple solutions. In this example we see that sometimes the problem is perfectly well behaved but if we are not careful we may think we find multiple solutions.

$$y' = 4y^{1/2}x \qquad y(0) = 4.$$

$$\int y^{-1/2} y' dx = \int 4x dx$$
$$2y^{1/2} = 2x^2 + C$$
$$y = (x^2 + C)^2$$

Now by plugging in the initial condition we find that,

$$4=(0^2+C)^2=C^2$$

At this point it is very tempting to claim that:

$$C=\pm 2$$

And we appear get two solutions,

$$y=(x^2+2)^2$$
$$y=(x^2-2)^2$$

If,

$$y=(x^2-2)^2, \text{ then}$$
$$y'=4x(x^2-2), \text{ and}$$
$$y^{1/2}=|x^2-2|.$$

Thus when we plug into $y'=4y^{1/2}x$, we are trying to equate,

$$4x(x^2-2) \text{ and } 4x|x^2-2|,$$

but notice the first equation is negative for $0<x<\sqrt{2}$ while the second equation is always positive.

This doesn't happen for the other function because we end up trying to compare $4x(x^2+2)y'=4x|x^2+2|$, since $x^2+2>0$ we can drop the absolute values in the second expression.

We introduced the ambiguity when we were solving for y by squaring both sides of the equation. We could have avoided this issue if we solved for C immediately after integrating, in which case we would have been working with the equation $2y^{1/2}=2x^2+C$ and:

$$2(4)^{1/2}=2(0)^2+C, \text{ so}$$
$$C=4$$

Plugging this back in and solving for y we get:

$$2y^{1/2}=2x^2+4$$
$$y^{1/2}=x^2+2$$
$$y=(x^2+2)^2$$

Examples of Separable Equations

Acceleration, Velocity and Position

The classic real world example of differential equations is the relationship between acceleration, velocity, and position.

$$a(t) = v'(t) = x''(t)$$

So if you're given an equation for acceleration, you can figure out both velocity and position.

Constant Acceleration

Lets say that accelearation is a constant, g (the acceleration due to gravity, or about 10 m s^{-2}. The initial velocity at t=0 is v_0. The initial position is x_0. Solve for v and x.

First, you need to solve for v. I.w.r.t. x.

$$v' = a = g$$

$$\int v' dv = \int g dt$$

$$v = gt + C$$

Now plug in to find C.

$$v_0 = g \times 0 + C$$

$$C = v_0$$

$$v(t) = gt + v_0$$

Now we solve for x.

$$x' = v = gt + v_0$$

$$\int x' dt = \int (gt + v_0) dt$$

$$x = \frac{1}{2} gt^2 + v_0 t + C$$

Again, now we solve for C.

$$x_0 = \frac{1}{2} g \times 0^2 + v_0 \times 0 + C$$

$$x_0 = C$$

$$x(t) = -\frac{1}{2} gt^2 + v_0 t + x_0$$

Resistive Medium

Lets say we're traveling through a medium that resists movement. In this medium, $a = -v^2$. Solve for v and x, given that the initial velocity was 10 m/s and the initial displacement was 0 m.

$$a = v' = -v^2$$

$$\frac{dv}{dt} = -v^2$$

$$\int \frac{1}{v^2} dv = \int -1 dt$$

$$-\frac{1}{v} = -t + C$$

$$v = \frac{1}{t + C}$$

Notice that as *t* increases, velocity decreases. This is what you'd expect if the medium was resisting your movement and slowing you down over time. Now substitute in the initial velocity at *t*=0.

$$10 = \frac{1}{0 + C}$$

$$C = \frac{1}{10}$$

$$v = \frac{1}{t + 0.1}$$

Now to solve for *x*.

$$v = x' = \frac{1}{t + 0.1}$$

$$\frac{dx}{dt} = \frac{1}{t + 0.1}$$

$$\frac{dx}{1} = \frac{dt}{t + 0.1}$$

$$\int 1 dx = \int \frac{1}{t + 0.1} dt$$

$$x = \ln(t + 0.1) + D$$

The position increases throughout, but it increases ever more slowly as time goes on. This is again what you'd expect for a medium resisting motion. Since velocity is never less than 0, we never stop going forward, but we go exponentially less far over time. Putting in our boundary conditions,

$$0 = \ln(0 + 0.1) + D$$

$$D = -\ln(0.1)$$

$$x = \ln(t + 0.1) - \ln(0.1) = \ln(10t + 1)$$

Exponential Growth and Decay

One of the most common differential equations in science is,

$$y' = ky$$

The solution to this is,

$$y = Ce^{kt}.$$

If *k* is positive, this is called exponential growth. If *k* is negative, its exponential decay. Both are used in science, for very different reasons.

Population Growth

Lets say we have a group of animals in the wild. We want to know how many animals there will be in *t* years. We know how many there are now. We also know the birth rate and death rate.

First, we need to figure out the rate of growth. If the birth rate is B, and the death rate is D, the total rate of change is (B-D). Since this is the rate, we need to multiply it by the current population to get the population growth. The final equation looks like,

$$\Delta P = (B - D)P$$

where P is the population. That looks like the equation for exponential growth, doesn't it? As a matter of fact, change the 'delta' to a differential, and it is. The growth factor is (B-D).

Example:

In a certain population of rabbits, the birth rate is 10%. The death rate is 15%. The initial population is 100. How many rabbits are there after 10 years? Will we always have rabbits?

From our solution to the exponential equation:

$$P = Ce^{(B-D)t}$$

$$100 = Ce^{(B-D)\times 0} = Ce^{0}$$

$$C = 100$$

$$P = 100e^{-0.05t}$$

$$P(10) = 100e^{-0.5} \approx 61$$

Unfortunately, we will not always have rabbits. Since the growth rate is negative, they will eventually go extinct. We will never actually hit 0, but in real life you can't have less than 1 rabbit. If we were measuring a continuous property instead of a discrete one, we would always have something, it would just get very small.

Radioactive Decay

Another situation for exponential growth is radioactive isotopes. If you have a sample of radioactive material, individual atoms will randomly decay or not decay. While you can't know exactly how many atoms decay and when, you do know the average rate of decay. Every λ years, half of the atoms left will decay. This period of time, λ, is called a half life. The *activity* of the sample (how many decays per second) is called A. Mathematically, this looks like

$$\frac{dA}{dt} = \lambda A$$

FIRST-ORDER ORDINARY DIFFERENTIAL EQUATION

Given a first-order ordinary differential equation,

$$\frac{dy}{dx} = F(x,y),$$

if $F(x,y)$ can be expressed using separation of variables as,

$$F(x,y) = X(x)Y(y),$$

then the equation can be expressed as,

$$\frac{dy}{Y(y)} = X(x)dx$$

and the equation can be solved by integrating both sides to obtain,

$$\int \frac{dy}{Y(y)} = \int X(x)dx.$$

Any first-order ODE of the form,

$$\frac{dy}{dx} + p(x)y = q(x)$$

can be solved by finding an integrating factor $\mu = \mu(x)$ such that,

$$\frac{d}{dx}(\mu y) = \mu \frac{dy}{dx} + y\frac{d\mu}{dx}$$
$$= \mu q(x)$$

Dividing through by μ y yields,

$$\frac{1}{y}\frac{dy}{dx} + \frac{1}{\mu}\frac{d\mu}{dx} = \frac{q(x)}{y}.$$

However, this condition enables us to explicitly determine the appropriate μ for arbitrary p and q. To accomplish this, take,

$$p(x) = \frac{1}{\mu}\frac{d\mu}{dx}$$

in the above equation, from which we recover the original equation $\frac{1}{y}\frac{dy}{dx} + \frac{1}{\mu}\frac{d\mu}{dx} = \frac{q(x)}{y}$. As required, in the form,

$$\frac{1}{y}\frac{dy}{dx} + p(x) = \frac{q(x)}{y}.$$

But we can integrate both sides of equation $p(x) = \frac{1}{\mu}\frac{d\mu}{dx}$ to obtain,

$$\int p(x)dx = \int \frac{d\mu}{\mu} = \ln\mu + c$$

$$\mu = e^{\int p(x)dx}.$$

Now integrating both sides of $\frac{1}{y}\frac{dy}{dx} + \frac{1}{\mu}\frac{d\mu}{dx} = \frac{q(x)}{y}$ gives,

$$\mu y = \int \mu q(x)dx + c$$

with μ now a known function), which can be solved for y to obtain,

$$y = \frac{\int \mu q(x)dx + c}{\mu} = \frac{\int e^{\int^{x} p(x')dx'} q(x)dx + c}{e^{\int^{x} p(x')dx'}},$$

where c is an arbitrary constant of integration.

Given an nth-order linear ODE with constant coefficients,

$$\frac{d^n y}{dx^n} + a_{(n-1)}\frac{d^{(n-1)}y}{dx^{(n-1)}} + \ldots + a_1\frac{dy}{dx} + a_0 y = Q(x),$$

first solve the characteristic equation obtained by writing,

$$y \equiv e^{rx}$$

and setting $Q(x) = 0$ to obtain the n complex roots,

$$r^n e^{rx} + a_{n-1}r^{n-1}e^{rx} + \ldots + a_1 r e^{rx} + a_0 e^{rx} = 0$$

$$r^n + a_{n-1}r^{n-1} + \ldots + a_1 r + a_0 = 0.$$

Factoring gives the roots r_i,

$$(r - r_1)(r - r_2)\ldots(r - r_n) = 0.$$

For a nonrepeated real root r, the corresponding solution is,

$$y = e^{(rx)}.$$

If a real root r is repeated k times, the solutions are degenerate and the linearly independent solutions are,

$$y = e^{rx}, y = xe^{rx}, \ldots, y = x^{k-1}e^{rx}.$$

Complex roots always come in complex conjugate pairs, $r_{\pm} = a \pm ib$. For nonrepeated complex roots, the solutions are,

$$y = e^{ax}\cos(bx), y = e^{ax}\sin(bx).$$

If the complex roots are repeated k times, the linearly independent solutions are,

$$y = e^{ax}\cos(bx), y = e^{ax}\sin(bx), \ldots, y = x^{k-1}e^{ax}\cos(bx), y = x^{k-1}e^{ax}\sin(bx).$$

Linearly combining solutions of the appropriate types with arbitrary multiplicative constants then gives the complete solution. If initial conditions are specified, the constants can be explicitly determined. For example, consider the sixth-order linear ODE,

$$(\tilde{D} - 1)(\tilde{D} - 2)^3(\tilde{D}^2 + \tilde{D} + 1)y = 0,$$

which has the characteristic equation,

$$(r-1)(r-2)^3(r^2 + r + 1) = 0.$$

The roots are 1, 2 (three times), and $(-1 \pm \sqrt{3}i)/2$, so the solution is,

$$y = Ae^{x} + Be^{2x} + Cxe^{2x} + Dx^2e^{2x} + E\,e^{-x/2}\cos\left(\frac{1}{2}\sqrt{3}\,x\right) + F\,e^{-x/2}\sin\left(\frac{1}{2}\sqrt{3}\,x\right).$$

If the original equation is nonhomogeneous ($Q(x) \neq 0$), now find the particular solution y^* by the method of variation of parameters. The general solution is then,

$$y(x) = \sum_{i=1}^{n} c_i y_i(x) + y^*(x),$$

where the solutions to the linear equations are $y_1(x), y_2(x), \ldots, y_n(x)$, and $y^*(x)$ is the particular solution.

SECOND-ORDER ORDINARY DIFFERENTIAL EQUATION

An ordinary differential equation of the form,

$$y'' + P(x)y' + Q(x)y = 0.$$

Such an equation has singularities for finite $x = x_0$ under the following conditions: (a) If either $P(x)$ or $Q(x)$ diverges as $x \to x_0$, but $(x - x_0)P(x)$ and $(x - x_0)^2\, Q(x)$ remain finite as $x \to x_0$, then x_0 is called a regular or nonessential singular point. (b) If $P(x)$ diverges faster than $(x - x_0)^{-1}$ so that $(x - x_0)P(x) \to \infty$ as $x \to x_0$, or $Q(x)$ diverges faster than $(x - x_0)^{-2}$ so that $(x - x_0)^2\, Q(x) \to \infty$ as $x \to x_0$, then is called an irregular or essential singularity.

Singularities of equation $y'' + P(x)y' + Q(x)y = 0$ at infinity are investigated by making the substitution $x \equiv z^{-1}$, so $dx = -z^{-2}dz$ giving,

$$\frac{dy}{dx} = -z^2 \frac{dy}{dx}$$

$$\frac{d^2y}{dx^2} = -z^2 \frac{d}{dz}\left(-z^2 \frac{dy}{dz}\right)$$

$$= -z^2\left(-2z\frac{dy}{dz} - z^2\frac{d^2y}{dz^2}\right)$$

$$= 2z^3\frac{dy}{dz} + z^4\frac{d^2y}{dz^2}.$$

Then $\frac{d^2y}{dx^2} = -z^2 \frac{d}{dz}\left(-z^2 \frac{dy}{dz}\right)$ becomes,

$$z^4\frac{d^2y}{dz^2} + [2z^3 - z^2P(z^{-1})]\frac{dy}{dz} + Q(z^{-1})y = 0.$$

Case (a): If

$$\alpha(z) = \frac{2z - P(z^{-1})}{z^2}$$

$$\beta(z) = \frac{Q(z^{-1})}{z^4}$$

remain finite at $x = \pm\infty$ $(z = 0)$, then the point is ordinary. Case (b): If either $\alpha(z)$ diverges no more rapidly than $1/z$ or $\beta(z)$ diverges no more rapidly than $1/z^2$, then the point is a regular singular point. Case (c): Otherwise, the point is an irregular singular point.

Morse and Feshbach give the canonical forms and solutions for second-order ordinary differential equations classified by types of singular points.

For special classes of linear second-order ordinary differential equations, variable coefficients can be transformed into constant coefficients. Given a second-order linear ODE with variable coefficients,

$$\frac{d^2y}{dx^2} + p(x)\frac{dy}{dx} + q(x)y = 0.$$

Define a function $z \equiv y(x)$,

$$\frac{dy}{dx} = \frac{dz}{dx}\frac{dy}{dz}$$

$$\frac{d^2y}{dx^2} = \left(\frac{dz}{dx}\right)^2 \frac{d^2y}{dz^2} + \frac{d^2z}{dx^2}\frac{dy}{dz}$$

$$\left(\frac{dz}{dx}\right)^2 \frac{d^2y}{dz^2} + \left[\frac{d^2z}{dx^2} + p(x)\frac{dz}{dx}\right]\frac{dy}{dz} + q(x)y = 0$$

$$\frac{d^2y}{dz^2} + \left[\frac{\frac{d^2z}{dx^2} + p(x)\frac{dz}{dx}}{\left(\frac{dz}{dx}\right)^2}\right]\frac{dy}{dz} + \left[\frac{q(x)}{\left(\frac{dz}{dx}\right)^2}\right]y$$

$$\equiv \frac{d^2y}{dz^2} + A\frac{dy}{dz} + By = 0$$

This will have constant coefficients if A and B are not functions of x. But we are free to set B to an arbitrary positive constant for $q(x) \geq 0$ by defining z as,

$$z = B^{(-1/2)}\int [q(x)]^{(1/2)} dx.$$

Then,

$$\frac{dz}{dx} = B^{-1/2}\left[q(x)\right]^{1/2}$$

$$\frac{d^2z}{dx^2} = \frac{1}{2}B^{-1/2}\left[q(x)\right]^{-1/2} q'(x),$$

and,

$$A = \frac{\frac{1}{2}B^{-1/2}\left[q(x)\right]^{-1/2} q'(x) + B^{-1/2} p(x)\left[q(x)\right]^{1/2}}{B^{-1}q(x)}$$

$$= \frac{q'(x) + 2p(x)q(x)}{2\left[q(x)\right]^{3/2}} B^{1/2}$$

Equation therefore becomes,

$$\frac{d^2y}{dz^2} + \frac{q(x) + 2\,p(x)\,q(x)}{2[q(x)]^{(3/2)}} B^{1/2}\frac{dy}{dz} + By = 0,$$

which has constant coefficients provided that,

$$A \equiv \frac{q'(x) + 2p(x)q(x)}{2[q(x)]^{3/2}} B^{1/2} = [\text{constant}].$$

Eliminating constants, this gives,

$$A' \equiv \frac{q'(x)+2p(x)\,q(x)}{[q(x)]^{3/2}} = [\text{constant}].$$

So for an ordinary differential equation in which A' is a constant, the solution is given by solving the second-order linear ODE with constant coefficients,

$$\frac{d^2y}{dz^2} + A\frac{dy}{dz} + By = 0$$

for z, where z is defined as above.

A linear second-order homogeneous differential equation of the general form,

$$y'' + P(x)y' + Q(x)y = 0$$

can be transformed into standard form,

$$z'' + q(x)z = 0$$

with the first-order term eliminated using the substitution,

$$\ln y \equiv \ln z - \frac{1}{2}\int P(x)dx.$$

Then,

$$\frac{y'}{y} = \frac{z'}{z} - \frac{1}{2}P(x)$$

$$\frac{yy'' - y'^2}{y^2} = \frac{zz'' - z'^2}{z^2} - \frac{1}{2}P'(x)$$

$$\frac{y''}{y} - \left(\frac{y'}{y}\right)^2 = \frac{z''}{z} - \frac{z'^2}{z^2} - \frac{1}{2}P'(x)$$

$$\frac{y''}{y} = \left[\frac{z'}{z} - \frac{1}{2}P(x)\right]^2 + \frac{z''}{z} - \frac{z'^2}{z^2} - \frac{1}{2}P'(x)$$

$$= \frac{z'^2}{z^2} - \frac{1}{2}P'(x) + \frac{1}{4}P^2(x) + \frac{z''}{z} - \frac{z'^2}{z^2} - \frac{1}{2}P'(x),$$

So,

$$\frac{y''}{y} + P(x)\frac{y'}{y} + Q(x) = -\frac{z'}{z}P(x) + \frac{1}{4}P^2(x) + \frac{z''}{z} - \frac{1}{2}P'(x) + P(x)\left[\frac{z'}{z} - \frac{1}{2}P(x)\right] + Q(x)$$

$$= \frac{z''}{z} - \frac{1}{2}P'(x) - \frac{1}{4}P^2(x) + Q(x) = 0$$

Therefore,

$$z''+\left[Q(x)-\frac{1}{2}P'(x)-\frac{1}{4}P^2(x)\right]z \equiv z''(x)+q(x)z=0,$$

where,

$$q(x)\equiv Q(x)-\frac{1}{2}P'(x)-\frac{1}{4}P^2(x).$$

If $Q(x)=0$, then the differential equation becomes,

$$y''+P(x)y'=0,$$

which can be solved by multiplying by,

$$\exp\left[\int^x P(x')dx'\right]$$

to obtain,

$$0=\frac{d}{dx}\left\{\exp\left[\int^x P(x')dx'\right]\frac{dy}{dx}\right\}$$

$$c_1=\exp\left[\int^x P(x')dx'\right]\frac{dy}{dx}$$

$$y=c_1\int^x \frac{dx}{\exp\left[\int^x P(x')dx'\right]}+c_2$$

For a nonhomogeneous second-order ordinary differential equation in which the x term does not appear in the function $f(x,y,y')$,

$$\frac{d^2y}{dx^2}=f(y,y'),$$

let $v\equiv y'$, then,

$$\frac{dv}{dx}=f(v,y)=\frac{dv}{dy}\frac{dy}{dx}=v\frac{dv}{dy}.$$

So the first-order ODE,

$$v\frac{dv}{dy}=f(y,v),$$

if linear, can be solved for v as a linear first-order ODE. Once the solution is known,

$$\frac{dy}{dx}=v(y)$$

$$\int \frac{dy}{v(y)} = \int dx$$

On the other hand, if y is missing from $f(x,y,y')$,

$$\frac{d^2y}{dx^2} = f(x,y'),$$

let $v \equiv y'$, then $v' = y''$, and the equation reduces to,

$$v' = f(x,v),$$

which, if linear, can be solved for v as a linear first-order ODE. Once the solution is known,

$$y = \int v(x)dx.$$

Nonhomogeneous ordinary differential equations can be solved if the general solution to the homogenous version is known, in which case variation of parameters can be used to find the particular solution. In particular, the particular solution $y^*(x)$ to a nonhomogeneous second-order ordinary differential equation,

$$y'' + p(x)y' + q(x)y = g(x)$$

can be found using variation of parameters to be given by the equation,

$$y^*(x) = -y_1(x)\int \frac{y_2(x)g(x)}{W(x)}dx + y_2(x)\int \frac{y_1(x)g(x)}{W(x)},$$

where $y_1(x)$ and $y_2(x)$ are the homogeneous solutions to the unforced equation,

$$y'' + p(x)y' + q(x)y = 0$$

and W(x) is the Wronskian of these two functions.

ABEL'S IDENTITY

In mathematics, Abel's identity (also called as Abel's Formula or Abel's differential equation identity) is an equation that expresses the Wronskian of two solutions of a homogeneous second-order linear ordinary differential equation in terms of a coefficient of the original differential equation. The relation can be generalised to *n*th-order linear ordinary differential equations. The identity is named after the Norwegian mathematician Niels Henrik Abel.

Since Abel's identity relates the different linearly independent solutions of the differential equation, it can be used to find one solution from the other. It provides useful identities relating the solutions, and is also useful as a part of other techniques such as the method of variation of parameters. It is especially useful for equations such as Bessel's equation where the solutions do not have a simple analytical form, because in such cases the Wronskian is difficult to compute directly.

A generalisation to first-order systems of homogeneous linear differential equations is given by Liouville's formula.

Consider a homogeneous linear second-order ordinary differential equation:

$$y'' + p(x)y' + q(x)y = 0$$

on an interval *I* of the real line with real- or complex-valued continuous functions *p* and *q*. Abel's identity states that the Wronskian $W = (y_1, y_2)$ of two real- or complex-valued solutions y_1 and y_2 of this differential equation, that is the function defined by the determinant.

$$W(y_1, y_2)(x) = \begin{vmatrix} y_1(x) & y_2(x) \\ y_1'(x) & y_2'(x) \end{vmatrix} = y_1(x)y_2'(x) - y_1'(x)y_2(x), \qquad x \in I,$$

satisfies the relation,

$$W(y_1, y_2)(x) = C\exp\left(-\int_{x_0}^{x} p(x)dx\right), \qquad x \in I,$$

for every point x_0 in *I*, where *C* is an arbitrary constant.

- In particular, the Wronskian $W(y_1, y_2)$ is either always the zero function or always different from zero with the same sign at every point x in I. In the latter case, the two solutions y_1 and y_2 are linearly independent.
- It is not necessary to assume that the second derivatives of the solutions y_1 and y_2 are continuous.
- Abel's theorem is particularly useful if $p(x) = 0$, because it implies that *W=const.*

Proof:

Differentiating the Wronskian using the product rule gives (writing W for $W(y_1, y_2)$ and omitting the argument *x* for brevity),

$$\begin{aligned} W' &= y_1'y_2' + y_1y_2'' - y_1''y_2 - y_1'y_2' \\ &= y_1y_2'' - y_1''y_2. \end{aligned}$$

Solving for y'' in the original differential equation yields:

$$y'' = -(py' + qy).$$

Substituting this result into the derivative of the Wronskian function to replace the second derivatives of y_1 and y_2 gives:

$$\begin{aligned} W' &= -y_1(py_2' + qy_2) + (py_1' + qy_1)y_2 \\ &= -p(y_1y_2' - y'_1y_2) \\ &= -pW. \end{aligned}$$

This is a first-order linear differential equation, and it remains to show that Abel's identity gives the unique solution, which attains the value $W(x_0)$ at x_0. Since the function p is continuous on I, it is bounded on every closed and bounded subinterval of I and therefore integrable, hence,

$$V(x) = W(x)\exp\left(\int_{x_0}^{x} p(\xi)d\xi\right), \qquad x \in I,$$

is a well-defined function. Differentiating both sides, using the product rule, the chain rule, the derivative of the exponential function and the fundamental theorem of calculus, one obtains,

$$V'(x) = (W'(x) + W(x)p(x))\exp\left(\int_{xo}^{x} p(\xi)d\xi\right) = 0, \qquad x \in I,$$

due to the differential equation for W. Therefore, V has to be constant on I, because otherwise we would obtain a contradiction to the mean value theorem (applied separately to the real and imaginary part in the complex-valued case). Since $V(x_0) = W(x_0)$, Abel's identity follows by solving the definition of V for $W(x)$.

Generalization

Consider a homogeneous linear nth-order ($n \geq 1$) ordinary differential equation:

$$y^{(n)} + p_{n-1}(x)y^{(n-1)} + \cdots + p_1(x)y' + p_0(x)y = 0,$$

on an interval I of the real line with a real- or complex-valued continuous function p_{n-1}. The generalisation of Abel's identity states that the Wronskian $W(y_1,\ldots,y_n)$ of n real- or complex-valued solutions $y_1,\ldots,y_n$ of this nth-order differential equation, that is the function defined by the determinant,

$$W(y_1,\ldots,y_n)(x) = \begin{vmatrix} y_1(x) & y_2(x) & \cdots & y_n(x) \\ y_1'(x) & y_2'(x) & \cdots & y_n'(x) \\ \vdots & \vdots & \ddots & \vdots \\ y_1^{(n-1)}(x) & y_2^{(n-1)}(x) & \cdots & y_n^{(n-1)}(x) \end{vmatrix}, \qquad x \in I,$$

satisfies the relation:

$$W(y_1,\ldots,y_n)(x) = W(y_1,\ldots,y_n)(x_0)\exp\left(-\int_{x_0}^{x} p_{n-1}(\xi)d\xi\right), \qquad x \in I,$$

for every point x_0 in I.

Direct Proof

For brevity, we write W for $W(y_1,\ldots,y_n)$ and omit the argument x. It suffices to show that the Wronskian solves the first-order linear differential equation,

$$W' = -p_{n-1}W,$$

because the remaining part of the proof then coincides with the one for the case $n=2$.

In the case $n=1$ we have $W=y_1$ and the differential equation for W coincides with the one for y_1. Therefore, assume $n\ge 2$ in the following.

The derivative of the Wronskian W is the derivative of the defining determinant. It follows from the Leibniz formula for determinants that this derivative can be calculated by differentiating every row separately, hence,

$$W' = \begin{vmatrix} y_1' & y_2' & \cdots & y_n' \\ y_1' & y_2' & \cdots & y_n' \\ y_1'' & y_2'' & \cdots & y_n'' \\ y_1''' & y_2''' & \cdots & y_n''' \\ \vdots & \vdots & \ddots & \vdots \\ y_1^{(n-1)} & y_2^{(n-1)} & \cdots & y_n^{(n-1)} \end{vmatrix} + \begin{vmatrix} y_1 & y_2 & \cdots & y_n \\ y_1'' & y_2'' & \cdots & y_n'' \\ y_1'' & y_2'' & \cdots & y_n'' \\ y_1''' & y_2''' & \cdots & y_n''' \\ \vdots & \vdots & \ddots & \vdots \\ y_1^{(n-1)} & y_2^{(n-1)} & \cdots & y_n^{(n-1)} \end{vmatrix}$$

$$+\cdots+ \begin{vmatrix} y_1 & y_2 & \cdots & y_n \\ y_1' & y_2' & \cdots & y_n' \\ \vdots & \vdots & \ddots & \vdots \\ y_1^{(n-3)} & y_2^{(n-3)} & \cdots & y_n^{(n-3)} \\ y_1^{(n-2)} & y_2^{(n-2)} & \cdots & y_n^{(n-2)} \\ y_1^{(n)} & y_2^{(n)} & \cdots & y_n^{(n)} \end{vmatrix}.$$

However, note that every determinant from the expansion contains a pair of identical rows, except the last one. Since determinants with linearly dependent rows are equal to 0, one is only left with the last one:

$$W' = \begin{vmatrix} y_1 & y_2 & \cdots & y_n \\ y_1' & y_2' & \cdots & y_n' \\ \vdots & \vdots & \ddots & \vdots \\ y_1^{(n-2)} & y_2^{(n-2)} & \cdots & y_n^{(n-2)} \\ y_1^{(n)} & y_2^{(n)} & \cdots & y_n^{(n)} \end{vmatrix}.$$

Since every y_i solves the ordinary differential equation, we have:

$$y_i^{(n)} + p_{n-2}y_i^{(n-2)} + \cdots + p_1 y_i' + p_0 y_i = -p_{n-1}y_i^{(n-1)}$$

for every $i \in \{1,\ldots,n\}$. Hence, adding to the last row of the above determinant p_0 times its first row, p_1 times its second row, and so on until p_{n-2} times its next to last row, the value of the determinant for the derivative of W is unchanged and we get,

$$W' = \begin{vmatrix} y_1 & y_2 & \cdots & y_n \\ y_1' & y_2' & \cdots & y_n' \\ \vdots & \vdots & \ddots & \vdots \\ y_1^{(n-2)} & y_2^{(n-2)} & \cdots & y_n^{(n-2)} \\ -p_{n-1}y_1^{(n-1)} & -p_{n-1}y_2^{(n-1)} & \cdots & -p_{n-1}y_n^{(n-1)} \end{vmatrix} = -p_{n-1}W.$$

Proof using Liouville's Formula

The solutions $y_1,\ldots,y_n$ form the square-matrix valued solution,

$$\Phi(x) = \begin{pmatrix} y_1(x) & y_2(x) & \cdots & y_n(x) \\ y_1'(x) & y_2'(x) & \cdots & y_n'(x) \\ \vdots & \vdots & \ddots & \vdots \\ y_1^{(n-2)}(x) & y_2^{(n-2)}(x) & \cdots & y_n^{(n-2)}(x) \\ y_1^{(n-1)}(x) & y_2^{(n-1)}(x) & \cdots & y_n^{(n-1)}(x) \end{pmatrix}, \qquad x \in I,$$

of the n -dimensional first-order system of homogeneous linear differential equations:

$$\begin{pmatrix} y' \\ y'' \\ \vdots \\ y^{(n-1)} \\ y^{(n)} \end{pmatrix} = \begin{pmatrix} 0 & 1 & 0 & \cdots & 0 \\ 0 & 0 & 1 & \cdots & 0 \\ \vdots & \vdots & \vdots & \ddots & \vdots \\ 0 & 0 & 0 & \cdots & 1 \\ -p_0(x) & -p_1(x) & -p_2(x) & \cdots & -p_{n-1}(x) \end{pmatrix} \begin{pmatrix} y \\ y' \\ \vdots \\ y^{(n-2)} \\ y^{(n-1)} \end{pmatrix}.$$

The trace of this matrix is $-p_{n-1}(x)$, hence Abel's identity follows directly from Liouville's formula.

ATKINSON–MINGARELLI THEOREM

In applied mathematics, the Atkinson–Mingarelli theorem, named after Frederick Valentine Atkinson and A. B. Mingarelli, concerns eigenvalues of certain Sturm–Liouville differential operators.

In the simplest of formulations let p, q, w be real-valued piecewise continuous functions defined on a closed bounded real interval, $I = [a, b]$. The function $w(x)$, which is sometimes denoted by $r(x)$, is called the "weight" or "density" function.

Consider the Sturm–Liouville differential equation:

$$-\frac{d}{dx}\left[p(x)\frac{dy}{dx}\right]+q(x)y=\lambda w(x)y,$$

where *y* is a function of the independent variable *x*. In this case, *y* is called a *solution* if it is continuously differentiable on (*a*,*b*) and (*p y'*)(*x*) is piecewise continuously differentiable and *y* satisfies the equation ($-\frac{d}{dx}\left[p(x)\frac{dy}{dx}\right]+q(x)y=\lambda w(x)y,$) at all except a finite number of points in (*a*,*b*). The unknown function *y* is typically required to satisfy some boundary conditions at *a* and *b*.

The boundary conditions under consideration here are usually called *separated boundary conditions* and they are of the form:

$$\alpha_1 y(a)+\alpha_2 y'(a)=0 \qquad (\alpha_1^2+\alpha_2^2>0),$$

$$\beta_1 y(b)+\beta_2 y'(b)=0 \qquad (\beta_1^2+\beta_2^2>0),$$

where the $\{\alpha_i,\beta_i\}$, $i=1, 2$ are real numbers.

Assume that *p*(*x*) has a finite number of sign changes and that the positive (resp. negative) part of the function *p*(*x*)/*w*(*x*) defined by $(w/p)_+(x)=\max\{w(x)/p(x),0\}$, (resp. $(w/p)_-(x)=\max\{-w(x)/p(x),0\}$) are not identically zero functions over I. Then the eigenvalue problem $-\frac{d}{dx}\left[p(x)\frac{dy}{dx}\right]+q(x)y=\lambda w(x)y,$

$$\alpha_1 y(a)+\alpha_2 y'(a)=0 \qquad (\alpha_1^2+\alpha_2^2>0),$$

$$\beta_1 y(b)+\beta_2 y'(b)=0 \qquad (\beta_1^2+\beta_2^2>0),$$

has an infinite number of real positive eigenvalues λ_i^+,

$$0<\lambda_1^+<\lambda_2^+<\lambda_3^+<\cdots<\lambda_n^+<\cdots\to\infty;$$

and an infinite number of negative eigenvalues λ_i^-,

$$0>\lambda_1^->\lambda_2^->\lambda_3^->\cdots>\lambda_n^->\cdots\to-\infty;$$

whose spectral asymptotics are given by their solution of Jörgens' Conjecture:

$$\lambda_n^+\sim\frac{n^2\pi^2}{\left(\int_a^b\sqrt{(w/p)_+(x)}\,dx\right)^2},\quad n\to\infty,$$

and

$$\lambda_n^- \sim \frac{-n^2\pi^2}{\left(\int_a^b \sqrt{(w/p)_-(x)}dx\right)^2}, \quad n \to \infty.$$

AUTONOMOUS SYSTEM

In mathematics, an autonomous system or autonomous differential equation is a system of ordinary differential equations which does not explicitly depend on the independent variable. When the variable is time, they are also called time-invariant systems.

Many laws in physics, where the independent variable is usually assumed to be time, are expressed as autonomous systems because it is assumed the laws of nature which hold now are identical to those for any point in the past or future.

Autonomous systems are closely related to dynamical systems. Any autonomous system can be transformed into a dynamical system and, using very weak assumptions, a dynamical system can be transformed into an autonomous system.

An autonomous system is a system of ordinary differential equations of the form:

$$\frac{d}{dt}x(t) = f(x(t))$$

where x takes values in n-dimensional Euclidean space and t is usually time.

It is distinguished from systems of differential equations of the form:

$$\frac{d}{dt}x(t) = g(x(t), t)$$

in which the law governing the rate of motion of a particle depends not only on the particle's location, but also on time; such systems are not autonomous.

Properties

Let $x_1(t)$ be a unique solution of the initial value problem for an autonomous system,

$$\frac{d}{dt}x(t) = f(x(t)), \quad x(0) = x_0.$$

Then $x_2(t) = x_1(t - t_0)$ solves,

$$\frac{d}{dt}x(t) = f(x(t)), \quad x(t_0) = x_0$$

Indeed, denoting $s = t - t_0$ we have $x_1(s) = x_2(t)$ and ds = dt, thus,

$$\frac{d}{dt}x_2(t) = \frac{d}{dt}x_1(t - t_0) = \frac{d}{ds}x_1(s) = f(x_1(s)) = f(x_2(t))$$

For the initial condition, the verification is trivial,

$$x_2(t_0) = x_1(t_0 - t_0) = x_1(0) = x_0$$

Qualitative Analysis

Autonomous systems can be analyzed qualitatively using the phase space; in the one-variable case, this is the phase line.

Solution Techniques

The following techniques apply to one-dimensional autonomous differential equations. Any one-dimensional equation of order n is equivalent to an n-dimensional first-order system, but not necessarily vice versa.

First Order

The first-order autonomous equation,

$$\frac{dx}{dt} = f(x)$$

is separable, so it can easily be solved by rearranging it into the integral form:

$$t + C = \int \frac{dx}{f(x)}$$

Second Order

The second-order autonomous equation,

$$\frac{d^2x}{dt^2} = f(x, x')$$

is more difficult, but it can be solved by introducing the new variable,

$$v = \frac{dx}{dt}$$

and expressing the second derivative of x (via the chain rule) as,

$$\frac{d^2x}{dt^2} = \frac{dv}{dt} = \frac{dx}{dt}\frac{dv}{dx} = v\frac{dv}{dx}$$

so that the original equation becomes:

$$v\frac{dv}{dx} = f(x, v)$$

which is a first order equation containing no reference to the independent variable t and if solved provides v as a function of x. Then, recalling the definition of v:

$$\frac{dx}{dt} = v(x) \Rightarrow t + C = \int \frac{dx}{v(x)}$$

which is an implicit solution.

Special Case: $x'' = f(x)$

The special case where f is independent of x',

$$\frac{d^2x}{dt^2} = f(x)$$

benefits from separate treatment. These types of equations are very common in classical mechanics because they are always Hamiltonian systems.

The idea is to make use of the identity (barring division by zero issues),

$$\frac{x}{dt} = \left(\frac{dt}{dx}\right)^{-1}$$

which follows from the chain rule. Note aside then that by inverting both sides of a first order autonomous system, one can immediately integrate with respect to x:

$$\frac{dx}{dt} = f(x) \Rightarrow \frac{dt}{dx} = \frac{1}{f(x)} \Rightarrow t + C = \int \frac{dx}{f(x)}$$

which is another way to view the separation of variables technique. A natural question is then: can we do something like this with higher order equations? The answer is yes for second order equations, but there's more work to do. The second derivative must be expressed as a derivative with respect to x instead of t:

$$\begin{aligned} \frac{d^2x}{dt^2} &= \frac{d}{dt}\left(\frac{dx}{dt}\right) = \frac{d}{dx}\left(\frac{dx}{dt}\right)\frac{dx}{dt} = \frac{d}{dx}\left(\left(\frac{dt}{dx}\right)^{-1}\right)\left(\frac{dt}{dx}\right)^{-1} = \\ &= -\left(\frac{dt}{dx}\right)^{-2}\frac{d^2t}{dx^2}\left(\frac{dt}{dx}\right)^{-1} = -\left(\frac{dt}{dx}\right)^{-3}\frac{d^2t}{dx^2} = \\ &= \frac{d}{dx}\left(\frac{1}{2}\left(\frac{dt}{dx}\right)^{-2}\right) \end{aligned}$$

To reemphasize: what's been accomplished is that the second derivative in t has been expressed as a derivative in x. The original second order equation may then finally be integrated:

$$\frac{d^2x}{dt^2} = f(x)$$

$$\frac{d}{dx}\left(\frac{1}{2}\left(\frac{dt}{dx}\right)^{-2}\right)=f(x)$$

$$\left(\frac{dt}{dx}\right)^{-2}=2\int f(x)dx+C_1$$

$$\frac{t}{dx}=\pm\frac{1}{\sqrt{2\int f(x)dx+C_1}}$$

$$t+C_2=\pm\int\frac{dx}{\sqrt{2\int f(x)dx+C_1}}$$

This is an implicit solution, and beyond that the greatest potential problem is inability to simplify the integrals, which implies difficulty or impossibility in evaluating the integration constants.

Special Case: $x'' = x'^n f(x)$

$$\frac{d^2x}{dt^2}=\left(\frac{dx}{dt}\right)^n f(x)$$

where n is some parameter not equal to two. This will work since the second derivative can be written in a form involving a power of x'. Rewriting the second derivative, rearranging, and expressing the left side as a derivative:

$$-\left(\frac{dt}{dx}\right)^{-3}\frac{d^2t}{dx^2}=\left(\frac{dt}{dx}\right)^{-n}f(x)$$

$$-\left(\frac{dt}{dx}\right)^{n-3}\frac{d^2t}{dx^2}=f(x)$$

$$\frac{d}{dx}\left(\frac{1}{2-n}\left(\frac{dt}{dx}\right)^{n-2}\right)=f(x)$$

$$\left(\frac{dt}{dx}\right)^{n-2}=(2-n)\int f(x)dx+C_1$$

$$t+C_2=\int\left((2-n)\int f(x)dx+C_1\right)^{\frac{1}{n-2}}dx$$

The right will carry +/- if n is even. The treatment must be different if $n=2$:

$$-\left(\frac{dt}{dx}\right)^{-1}\frac{d^2t}{dx^2}=f(x)$$

$$-\frac{d}{dx}(\ln(\frac{dt}{dx})) = f(x)$$

$$\frac{dt}{dx} = C_1 e^{-\int f(x)dx}$$

$$t + C_2 = C_1 \int e^{-\int f(x)dx} dx$$

Higher Orders

There is no analogous method for solving third- or higher-order autonomous equations. Such equations can only be solved exactly if they happen to have some other simplifying property, for instance linearity or dependence of the right side of the equation on the dependent variable only (i.e., not its derivatives). This should not be surprising, considering that nonlinear autonomous systems in three dimensions can produce truly chaotic behavior such as the Lorenz attractor and the Rössler attractor.

With this mentality, it also isn't too surprising that general non-autonomous equations of second order can't be solved explicitly, since these can also be chaotic (an example of this is a periodically forced pendulum).

BERNOULLI DIFFERENTIAL EQUATION

In mathematics, an ordinary differential equation of the form,

$$y' + P(x)y = Q(x)y^n$$

is called a Bernoulli differential equation where n is any real number other than 0 or 1. It is named after Jacob Bernoulli, who discussed it in 1695. Bernoulli equations are special because they are nonlinear differential equations with known exact solutions. A famous special case of the Bernoulli equation is the logistic differential equation.

Transformation to a Linear Differential Equation

When $n = 0$, the differential equation is linear. When $n = 1$, it is separable. In these cases, standard techniques for solving equations of those forms can be applied. For $n \neq 0$ and $n \neq 1$, the substitution $u = y^{1-n}$ reduces any Bernoulli equation to a linear differential equation. For example, in the case $n = 2$, making the substitution $u = y^{-1}$ in the differential equation $\frac{dy}{dx} + \frac{1}{x}y = xy^2$ produces the equation $\frac{du}{dx} - \frac{1}{x}u = -x$, which is a linear differential equation.

Solution:

Let $x_0 \in (a,b)$ and

$$\begin{cases} z:(a,b) \to (0,\infty), & \text{if } \alpha \in \mathbb{R} \setminus \{1,2\}, \\ z:(a,b) \to \mathbb{R} \setminus \{0\}, & \text{if } \alpha = 2, \end{cases}$$

be a solution of the linear differential equation:

$$z'(x) = (1-\alpha)P(x)z(x) + (1-\alpha)Q(x).$$

Then we have that $y(x) := [z(x)]^{\frac{1}{1-\alpha}}$ is a solution of

$$y'(x) = P(x)y(x) + Q(x)y^{\alpha}(x), \, y(x_0) = y_0 := [z(x_0)]^{\frac{1}{1-\alpha}}.$$

And for every such differential equation, for all $\acute{a} > 0$ we have $y \equiv 0$ as solution for $y_0 = 0$.

Example:

Consider the Bernoulli equation:

$$y' - \frac{2y}{x} = -x^2y^2$$

The constant function $y = 0$ is a solution. Division by y^2 yields,

$$y'y^{-2} - \frac{2}{x}y^{-1} = -x^2$$

Changing variables gives the equations,

$$w = \frac{1}{y}$$

$$w' = \frac{-y'}{y^2}.$$

$$-w' - \frac{2}{x}w = -x^2$$

$$w' + \frac{2}{x}w = x^2$$

which can be solved using the integrating factor,

$$M(x) = e^{2\int \frac{1}{x}dx} = e^{2\ln x} = x^2.$$

Multiplying by $M(x)$,

$$w'x^2 + 2xw = x^4,$$

The left side is the derivative of wx^2. Integrating both sides with respect to x results in the equations,

$$\int w'x^2 + 2xw dx = \int x^4 dx$$

$$wx^2 = \frac{1}{5}x^5 + C$$

$$\frac{1}{y}x^2 = \frac{1}{5}x^5 + C$$

The solution for y is,

$$y = \frac{x^2}{\frac{1}{5}x^5 + C}.$$

EXACT DIFFERENTIAL EQUATION

Definition of an Exact Differential Equation

The equation $M(x,y) + N(x,y)dy = 0$ is an exact differential equation if there exists a function f of two variables x and y having continuous partial derivatives such that,

$$f_x(x,y) = M(x,y) \quad \text{and} \quad f_y(x,y) = N(x,y)$$

The general solution of the equation is f(x, y)= C.

You know that if f has continuous second partials, then,

$$\frac{\partial M}{\partial y} = \frac{\partial^2 f}{\partial y \partial x} = \frac{\partial^2 f}{\partial x \partial y} = \frac{\partial N}{\partial x}.$$

This suggests the following test for exactness.

Let M and N have continuous partial derivatives on an open disc R. The differential equation $M(x,y)dx + N(x,y)dy = 0$ is exact if and only if,

$$\frac{\partial M}{\partial y} = \frac{\partial N}{\partial x}$$

Exactness is a fragile condition in the sense that seemingly minor alterations in an exact equation can destroy its exactness. This is demonstrated in the following example.

Testing for Exactness

The differential equation $\left(xy^2+x\right)dx+yx^2dy=0$ is exact because,

$$\frac{\partial M}{\partial y}=\frac{\partial}{\partial y}\left[xy^2+x\right]=2xy \quad \text{and} \quad \frac{\partial N}{\partial x}=\frac{\partial}{\partial x}\left[yx^2\right]=2xy.$$

But the equation $\left(y^2+1\right)dx+xy\,dy=0$ is not exact, even though it is obtained by dividing both sides of the first equation by x.

The differential equation $\cos y\,dx+\left(y^2-x\sin y\right)dy=0$ is exact because,

$$\frac{\partial M}{\partial y}=\frac{\partial}{\partial y}[\cos y]=-\sin y \qquad \text{and} \quad \frac{\partial N}{\partial x}=\frac{\partial}{\partial x}\left[y^2-x\sin y\right]=-\sin y$$

But the equation $\cos y\,dx+\left(y^2-x\sin y\right)dy=0$ is not exact, even though it differs from the first equation only by a single sign.

Note that the test for exactness of $M(x,y)dx+N(x,y)dy=0$ is the same as the test for determining whether $F(x,y)=M(x,y)i+N(x,y)j$ is the gradient of a potential function. This means that a general solution $f(x,y)=C$ to an exact differential equation can be found by the method used to find a potential function for a conservative vector field.

Solving an Exact Differential Equation

Solve the differential equation $\left(2xy-3x^2\right)dx+\left(x^2-2y\right)dy=0.$

Solution: The given differential equation is exact because,

$$\frac{\partial M}{\partial y}=\frac{\partial}{\partial y}\left[2xy-3x^2\right]=2x\frac{\partial N}{\partial x}=\frac{\partial}{\partial x}\left[x^2-2y\right]$$

The general solution, $f(x,y)=C$, is given by,

$$\begin{aligned} f(x,y)&=\int M(x,y)dx \\ &=\int\left(2xy-3x^2\right)dx=x^2y-x^3+g(y) \end{aligned}$$

You determined $g(y)$ by integrating $N(x,y)$ with respect to y and reconciling the two expressions for $f(x,y)$. An alternative method is to partially differentiate this version of $f(x,y)$ with respect to y and compare the result with $N(x,y)$. In other words,

$$f_y(x,y)=\frac{\partial}{\partial y}\left[x^2y-x^3+g(y)\right]=x^2+g'(y)=\overbrace{x^2-2y}^{N(x,y)}.$$

$$\uparrow \qquad \uparrow$$

$$g'(y)=-2y$$

Thus, $g'(y)=-2y$ and it follows that $g(y)=-y^2+C_1$. Therefore,

$$f(x,y)=x^2y-x^3-y^2+C_1$$

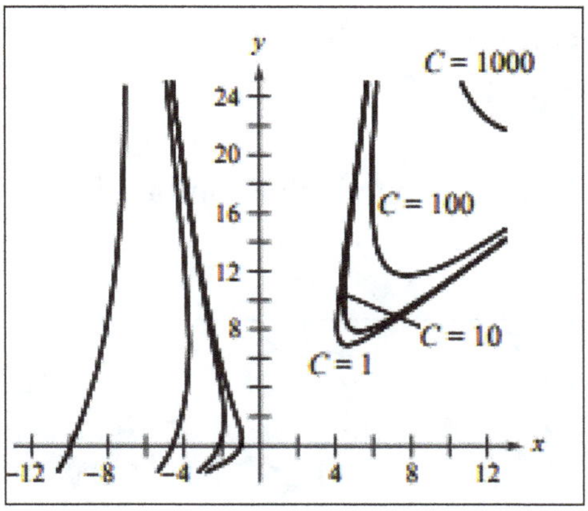

and the general solution is $x^2y-x^3-y^2=C$. Figure shows the solution curves that correspond to C=1, 10, 100, and 1000.

Example:

Find the particular solution of,

$$\left(\cos x - x\sin x + y^2\right)dx + 2xy\,dy = 0$$

that satisfies the initial condition $y=1$ when $x=\pi$.

Solution: The differential equation is exact because,

$$\overbrace{\frac{\partial}{\partial y}\left[\cos x - x\sin x + y^2\right]}^{\frac{\partial M}{\partial y}} = 2y = \overbrace{\frac{\partial}{\partial x}[2xy]}^{\frac{\partial N}{\partial x}}$$

Because $N(x,y)$ is simpler than $M(x,y)$ it is better to begin by integrating $N(x,y)$.

$$f(x,y)=\int N(x,y)dy=\int 2xy\,dy=xy^2+g(x)$$

$$f_x(x,y)=\frac{\partial}{\partial x}\left[xy^2+g(y)\right]=y^2+g'(x)=\overbrace{\cos x - x\sin x + y^2}^{M(x,y)}$$

$$\uparrow \qquad \uparrow$$

$$g'(x)=\cos x - x\sin x$$

Thus, $g'(x)=\cos x - x\sin x$ and

$$g(x)=\int(\cos x - x)\sin dx$$
$$=x\cos x+C$$

which implies that $f(x,y) = xy^2 + x\cos x + C_1$, and the general solution is:

$$xy^2 + x \cos x = C.$$

Applying the given initial condition produces,

$$\pi(1)^2 + \pi\cos\pi = c$$

which implies that C=0. Hence, the particular solution is,

$$xy^2 + x \cos x = 0$$

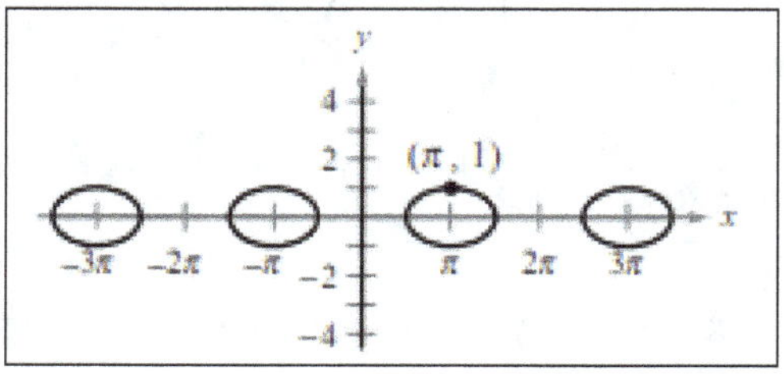

The graph of the particular solution is shown in figure. Notice that the graph consists of two parts: the ovals are given by $y^2 + x \cos x = 0$ and the y-axis is given by x=0.

Note that $z = f(x,y) = xy^2 + x\cos x$ if the total differential of z is given by,

$$\begin{aligned} dz &= f_x(x,y)\ dx + f_y(x,y)\ dy \\ &= (\cos x - x \sin x + y^2)\ dx + 2xy\ dy \\ &= M(x, y)\ dx\ 1\ N(x, y) dy \end{aligned}$$

In other words, $M dx + N dy = 0$ is called an exact differential equation because $M\ dx + N\ dy$ is exactly the differential of $f(x, y)$.

References

- Differential-Equation, First-Order-Ordinary: mathworld.wolfram.com, Retrieved 25 March, 2020
- Teschl, Gerald (2012). Ordinary Differential Equations and Dynamical Systems. Providence: American Mathematical Society. ISBN 978-0-8218-8328-0.
- Differential-Equation, Second-Order-Ordinary: mathworld.wolfram.com, Retrieved 25 March, 2020
- Hairer, Ernst; Nørsett, Syvert Paul; Wanner, Gerhard (1993), Solving ordinary differential equations I: Nonstiff problems, Berlin, New York: Springer-Verlag, ISBN 978-3-540-56670-0

3

Linear and Non-Linear Differential Equations

Any differential equation which is defined by a linear polynomial in the unknown function is known as a linear differential equation. A differential equation that is not a linear equation in the unknown function and its derivatives is termed as non-linear differential equation. This chapter has been carefully written to provide an easy understanding of linear and non-linear differential equations.

A general linear differential equation of order n, in the dependent variable y and the independent variable x, is an equation that can be expressed in the form:

$$a_0(x)\frac{d^n y}{dx^n}+a_1(x)\frac{d^{n-1} y}{dx^{n-1}}+.....+a_{n-1}(x)\frac{dy}{dx}+a_n(x)y=b(x)$$

where a_0 is not identically 0.

Properties of a General Linear Differential Equation

They possess the following properties as follows:

- The function y and its derivatives occur in the equation up to the first degree only.
- No products of y and/or any of its derivatives are present.
- No transcendental functions – (trigonometric or logarithmic etc) of y or any of its derivatives occur.

A linear differential equation of the first order is a differential equation that involves only the function y and its first derivative. Such equations are physically suitable for describing various linear phenomena in biology, economics, population dynamics, and physics.

LINEAR FIRST ORDER DIFFERENTIAL EQUATIONS

A linear first order equation is one that can be reduced to a general form:

$$\frac{dy}{dx}+p(x)y=Q(x)$$

where P(x) and Q(x) are continuous functions in the domain of validity of the differential equation. If P(x) or Q(x) is equal to 0, the differential equation can be reduced to a variables separable form which can be easily solved. You can check this for yourselves.

Integrating Factor

To find the solution of the linear first order differential equation as defined above, we must introduce the concept of an integrating factor. An integrating factor is a term, which when multiplied by an expression, converts it to an exact differential i.e. a function which is the derivative of another function. For example – given an expression,

$$y-x\frac{dy}{dx}$$

If we multiply this expression by $\frac{1}{x^2}$, note that it becomes an exact differential.

$$\frac{1}{x^2}\cdot\left(y-x\frac{dy}{dx}\right)$$

$$=\frac{d}{dx}\left(\frac{y}{x}\right)$$

Solving the Linear First Order Differential Equation

For the general first order linear differential equation, we assume that an integrating factor, that is only a function of x, exists. Let us call it η(x). Then this integrating factor, if multiplied by the expression in the differential equation should reduce the expression to an exact differential.

$$\frac{dy}{dx}+P(x)y=Q(x)$$

$$\eta(x)\frac{dy}{dx}+\eta(x)p(x)y=\eta(x)Q(x)$$

On inspection, note that if we put $\eta(x)P(x) = \eta'(x)$, we can easily get an exact differential on the left hand side.

Product Rule of Differentiation

$$\eta(x)\frac{dy}{dx}+\eta'(x)y=\eta(x)Q(x)$$

$$\frac{d}{dx}(\eta(x)y)=\eta(x)Q(x)$$

Now integrate on both sides with respect to x:

$$\int\frac{d}{dx}(\eta(x)y)dx=\int\eta(x)Q(x)dx$$

$$\eta(x)y+c=\int\eta(x)Q(x)dx$$

$$y=\frac{\int\eta(x)Q(x)dx-c}{\eta(x)}$$

Thus, we have got our general solution in the terms of our integrating function $\eta(x)$. You might worry that $\eta(x)$ is unknown to us. But on the contrary, it is very easy to compute in terms of other known quantities. If you remember, we had used the condition $\eta(x)P(x) = \eta'(x)$ to get the above solution. This is actually a variables separable equation which can be easily solved for $\eta(x)$ in the following way:

$$\eta(x)P(x)=\eta'(x)$$

$$P(x)=\frac{\eta'(x)}{\eta(x)}$$

$$\int P(x)dx=\int\frac{\eta'(x)}{\eta(x)}dx$$

$$\int P(x)dx=\ln(\eta(x))+k$$

$$\eta(x)=e^{\int P(x)dx-k}$$

$$\eta(x)=Ke^{\int P(x)dx}\text{, where } K = e^{-k}$$

This expresses the integrating function $\eta(x)$ in terms of an arbitrary constant K, and the known function P(x). Let us now put this back into our solution of the differential equation.

$$y=\frac{\int\eta(x)Q(x)dx-c}{\eta(x)}$$

$$y=\frac{K\int e^{\int P(x)dx}Q(x)dx-c}{Ke^{\int P(x)dx}}$$

$$y=\frac{\int e^{\int P(x)dx}Q(x)dx-\frac{c}{K}}{e^{\int P(x)dx}}$$

Let us just denote the arbitrary constant $\frac{c}{K}$ as C which is simply another arbitrary constant. In a concise form, we can then write the differential equation along with its solution as:

$$\frac{dy}{dx}+p(x)y=Q(x)$$

has the solution, $y=\frac{\int \eta(x)Q(x)dx-c}{\eta(x)}$

where $\eta(x)=e^{\int P(x)dx}$.

Thus, if we express the given differential equation in the standard form, we can now easily write down the solution using the above formula. Look at the solved examples below that make the process clearer.

Equations Reducible to Linear Form: Bernoulli's Equation

A differential equation of the form:

$$y'+p(x)y=g(x)y^{\alpha}$$

where a is a Real Number, is known as the Bernoulli's Equation. If a = 0, or a =1, it is a straightforward Linear Differential Equation to solve. However, for other values of a, the following method reduces the equation to a linear form:

- Introduce a new dependent variable u(x) such that:

  $$u(x)=[y(x)]^{1-a}$$

- The derivative of u can be found out as:

  $$u'=(1-a)y^{-a}y'$$

- From the differential equation, the value of y' can be substituted in the expression for u':

  $$\begin{aligned}u'&=(1-a)y^{-a}(gy^{a}-py)\\&=(1-a)(g-py^{1-a})\\&=(1-a)(g-pu)\end{aligned}$$

- The expression for the derivative of u can be rearranged to get:

  $$u'+(1-a)pu=(1-a)g$$

which is now an equation in the linear form with P(x) = (1 – a)pu, and Q(x) = (1 – a) g. This can then be solved for u(x) in a manner similar to the one for solving Linear Differential Equations. In the end, one must remember to back-substitute and get the final solution i.e. y(x).

Example: Solve

$$x\frac{dy}{dx}-2y=x^3\cos 4x$$

Solution: The first step is to convert the given differential equation to the standard form $\frac{dy}{dx}+P(x)y=Q(x)$. On rearranging, we can obtain the given differential equation as:

$$\frac{dy}{dx}-\frac{2}{x}y=x^2\cos 4x$$

On comparing with the standard form, we get $P(x)=-\frac{2}{x}$ and $Q(x)=x^2\cos 4x$. We should now compute the integrating factor first, then proceed towards the general solution. From the formula $\eta(x)=e^{\int P(x)dx}$, we can work out the integrating factor here as:

$$\eta(x)=e^{\int(-\frac{2}{x})dx}$$

$$=e^{-2\ln x}$$

$$=e^{\ln(x^{-2})}=x^{-2}$$

Now we can use this in the formula that we had derived for the general solution.

$$y=\frac{\int\eta(x)Q(x)dx-c}{\eta(x)}$$

$$y=\frac{\int\eta(x)Q(x)dx-C}{\eta(x)}$$

$$y=x^2\left(\frac{\sin 4x}{4}-c\right)$$

Solve the Bernoulli's Equation

$y'=Ay-By^2$,where A and B are constants.

Solution: In the standard form, the equation can be rewritten as:

$$y'-Ay=-By^2$$

Clearly, a = 2 here, and we must go ahead with the substitution $u = y^{1-a} = y^{-1}$. On differentiating u, and substituting the value of y' from the differential equation, we get:

$$\begin{aligned} u' &= -y^{-2}y' \\ &= -y^{-2}(Ay - By^2) \end{aligned}$$

or

$$\begin{aligned} &= B - Ay^{-1} \\ &= B - Au \end{aligned}$$

On rearranging, we obtain our linear differential equation in u as:

$$u' + Au = B$$

The general solution then is:

$$u = \frac{\int \eta(x)Bdx - C}{\eta(x)}$$

where:

$$\eta(x) = e^{\int Adx} = e^{Ax}$$

Thus,

$$u = \frac{B\int e^{Ax}dx - C}{e^{Ax}}$$

$$= \frac{B}{A} - Ce^{-Ax}$$

Finally back-substituting for y, we get:

$$y = \frac{1}{u} = \frac{1}{\frac{B}{A} - Ce^{-Ax}}$$

where C is an arbitrary constant, and A and B are known constants. Note directly from the given equation that y(x) = 0 for all x, is also a solution. It is a singular solution since it can't be obtained from the general solution.

SECOND ORDER LINEAR DIFFERENTIAL EQUATIONS

Second order linear equations with constant coefficients; Fundamental solutions; Wronskian; Existence and Uniqueness of solutions; the characteristic equation; solutions of homogeneous linear equations; reduction of order; Euler equations.

$$y'' + p(t)\, y' + q(t)\, y = g(t).$$

Homogeneous Equations: If g(t) = 0, then the equation above becomes:

$$y'' + p(t)\, y' + q(t)\, y = 0.$$

It is called a homogeneous equation. Otherwise, the equation is nonhomogeneous (or inhomogeneous).

Trivial Solution: For the homogeneous equation above, note that the function y(t) = 0 always satisfies the given equation, regardless what p(t) and q(t) are. This constant zero solution is called the trivial solution of such an equation.

Second Order Linear Homogeneous Differential Equations with Constant Coefficients

For the most part, we will only learn how to solve second order linear equation with constant coefficients (that is, when *p(t)* and *q(t)* are constants). Since a homogeneous equation is easier to solve compares to its nonhomogeneous counterpart, we start with second order linear homogeneous equations that contain constant coefficients only:

$$a\, y'' + b\, y' + c\, y = 0.$$

Where a, b, and c are constants, $a \neq 0$.

A very simple instance of such type of equations is,

$$y'' - y = 0.$$

The equation's solution is any function satisfying the equality $y'' = y$. Obviously y1 = e t is a solution, and so is any constant multiple of it, $C_1 e^t$. Not as obvious, but still easy to see, is that $y_2 = e^{-t}$ is another solution (and so is any function of the form $C_2 e^{-t}$).

It can be easily verified that any function of the form,

$$y = C_1 e^t + C_2 e^{-t}$$

will satisfy the equation. In fact, this is the general solution of the above differential equation.

Principle of Superposition: If y1 and y2 are any two solutions of the homogeneous equation,

$$y''+p(t)\,y'+q(t)\,y=0.$$

Then any function of the form $y = C_1 y_1 + C_2 y_2$ is also a solution of the equation, for any pair of constants C_1 and C_2.

That is, for a homogeneous linear equation, any multiple of a solution is again a solution; any sum/difference of two solutions is again a solution; and the sum/difference of the multiples of any two solutions is again a solution. (This principle holds true for a homogeneous linear equation of any order; it is not a property limited only to a second order equation. It, however, does not hold, in general, for solutions of a nonhomogeneous linear equation.)

However, while the general solution of $y''+p(t)\,y'+q(t)\,y=0$. Will always be in the form of $C_1 y_1 + C_2 y_2$, where y_1 and y_2 are some solutions of the equation, the converse is not always true. Not every pair of solutions y_1 and y_2 could be used to give a general solution in the form $y = C_1 y_1 + C_2 y_2$. We shall see shortly the exact condition that y_1 and y_2 must satisfy that would give us a general solution of this form.

Fact: The general solution of a second order equation contains two arbitrary constants /coefficients. To find a particular solution, therefore, requires two initial values. The initial conditions for a second order equation will appear in the form:

$$y(t_0)= y_0,\text{ and } y'(t_0)= y'_0.$$

Question: Just by inspection, can you think of two (or more) functions that satisfy the equation $y''+4y=0$? (A solution of this equation is a function φ such that $\varphi''=-4\varphi$.)

Example: Find the general solution of,

$$y''-5y'=0.$$

There is no need to "guess" an answer here. We actually know a way to solve the equation already. Observe that if we let $u=y'$, then $u' = y''$. Substitute them into the equation and we get a new equation:

$$u' - 5u = 0.$$

This is a first order linear equation with p(t) = −5 and g(t) = 0.

The integrating factor is $\mu=e^{-5t}$.

$$u(t)=\mu\frac{1}{(t)}\left(\int\mu(t)g(t)dt\right)=e^{5t}\left(\int 0dt\right)=e^{5t}(C)=Ce^{5t}$$

The actual solution y is given by the relation u = y′, and can be found by integration:

$$y(t)=\int u(t)dt=\int Ce^{5t}\,dt=\frac{C}{5}e^{5t}+C_2=C_1e^{5t}+C_2.$$

The method used in the above example can be used to solve any second order linear equation of the form $y''+\ p(t)\ y'=g(t)$, regardless whether its coefficients are constant or nonconstant, or it is a homogeneous equation or nonhomogeneous.

Equations of Nonconstant Coefficients with Missing y-term

If the y-term (that is, the dependent variable term) is missing in a second order linear equation, then the equation can be readily converted into a first order linear equation and solved using the integrating factor method.

Example:

$$ty''+4\,y'=t^2$$

The standard form is,

$$y''+\frac{4}{t}y'=t.$$

Substitute:

$$u'+\frac{4}{t}u=t \ \rightarrow\ p(t)=\frac{4}{t},g(t)=t$$

Integrating factor is,

$$\mu=t^4$$

$$u(t)=\frac{1}{t^4}\left(\int t^5dt\right)=t^4\left(\frac{1}{6}t^6=C\right)=\frac{1}{6}t^2+Ct^{-4}$$

Finally,

$$y(t)=\int u(t)dt=\frac{1}{18}t^3-\frac{C}{3}t^{-3}+C_2=\frac{1}{18}t^3+C_1t^3+C_2$$

Notice the above solution is not in the form of $y = C_1 y_1 + C_2 y_2$. There is nothing wrong with this, because this equation is not homogeneous. The general solution of a nonhomogeneous linear equation has a slightly different form. We will learn about the solutions of nonhomogeneous linear equations a bit later.

In general, given a second order linear equation with the y-term missing,

$$y'' + p(t)y' = g(t),$$

we can solve it by the substitutions u = y′ and u′ = y″ to change the equation to a first order linear equation. Use the integrating factor method to solve for u, and then integrate u to find y. That is:

Substitute: $u' + p(t)u = g(t)$

Integrating factor: $\mu(t) = e^{\int p(t)dt}$

Solve for u: $u(t) = \dfrac{\int \mu(t)g(t)dt(+C)}{\mu(t)}$

Integrate: $y(t) = \int u(t)dt$

This method works regardless whether the coefficients are constant or nonconstant, or if the equation is nonhomogeneous.

The Characteristic Polynomial

Back to the subject of the second order linear homogeneous equations with constant coefficients:

$$ay'' + by' + cy = 0, \quad a \neq 0. \quad (*)$$

We have seen a few examples of such an equation. In all cases the solutions consist of exponential functions, or terms that could be rewritten into exponential functions† . With this fact in mind, let us derive a (very simple, as it turns out) method to solve equations of this type. We will start with the assumption that there are indeed some exponential functions of unknown exponents that would satisfy any equation of the above form. We will then devise a way to find the specific exponents that would give us the solution.

Let y = e^{rt} be a solution of (*), for some as-yet-unknown constant r. Substitute y, y′ = re^{rt}, and y″ = r^2e^{rt} into (*), we get:

$$e^{rt}\left(ar^2 + brc\right) = 0.$$

Since e^{rt} is never zero, the above equation is satisfied (and therefore y = e^{rt} is a solution of (*)) if and only if a r^2 + br + c = 0. Notice that the expression a r^2 + br + c is a quadratic polynomial with r as the unknown. It is always solvable, with roots given by the quadratic formula. Hence, we can always solve a second order linear homogeneous equation with constant coefficients (*).

This polynomial, $ar^2 + br + c$, is called the characteristic polynomial of the differential equation (*). The equation,

$$ar^2 + br + c = 0$$

is called the characteristic equation of (*). Each and every root, sometimes called a characteristic root, r, of the characteristic polynomial gives rise to a solution $y = e^{rt}$ of (*).

We will take a more detailed look of the 3 possible cases of the solutions thusly found:

- (When $b^2 - 4ac > 0$) There are two distinct real roots r_1, r_2.
- (When $b^2 - 4ac < 0$) There are two complex conjugate roots $r = \lambda \pm \mu i$.
- (When $b^2 - 4ac = 0$) There is one repeated real root r.

There is no need to put the equation in its standard form when solving it using the characteristic equation method. The roots of the characteristic equation remain the same regardless whether the leading coefficient is 1 or not.

Case 1: Two Distinct Real Roots

When $b^2 - 4ac > 0$, the characteristic polynomial have two distinct real roots r_1, r_2. They give two distinct‡ solutions $y_1 = e^{r_1 t}$ and $y_2 = e^{r_2 t}$. Therefore, a general solution of (*) is:

$$y = C_1 y_1 + C_2 y_2 = C_1 e^{r_1 t} + C_2 e^{r_2 t}.$$

Example: $y'' + 5y' + 4y = 0$

The characteristic equation is $r^2 + 5r + 4 = (r+1)(r+4) = 0$, the roots of the polynomial are $r = -1$ and -4. The general solution is then,

$$y = C_1 e^{-t} + C_2 e^{-4t}.$$

Suppose there are initial conditions $y(0) = 1$, $y'(0) = -7$. A unique particular solution can be found by solving for C_1 and C_2 using the initial conditions. First we need to calculate $y' = -C1e^{-t} - 4C_2 e^{-4t}$, then apply the initial values:

$$1 = y(0) = C_1 e^0 + C_2 e^0 = C_1 + C_2$$

$$-7 = y'(0) = -C_1 e^0 - 4C_2 e^0 = -C_1 - 4C_2$$

The solution is $C1 = -1$, and $C_2 = 2 \quad \rightarrow \quad y = -e^{-t} + 2e^{-4t}$.

Suppose the initial conditions are instead $y(10000) = 1$, $y'(10000) = -7$. How would the new t_0 change the particular solution?

Apply the initial conditions as before, and we see there is a little complication. Namely, the simultaneous system of 2 equations that we have to solve in order to find C_1 and C_2 now comes with rather inconvenient irrational coefficients:

$$1 = y(10000) = C_1 e^{-10000} + C_2 e^{-40000}$$
$$-7 = y'(10000) = -C_1 e^{-10000} - 4C_2 e^{-40000}$$

With some good bookkeeping, systems like this can be solved the usual way. However, there is an easier method to simplify the inconvenient coefficients. The idea is translation (or time-shift). What we will do is to first construct a new coordinate axis, say $\check{T}$-axis. The two coordinate-axes are related by the equation $\check{T} = t - t_0$ (Therefore, when $t = t_0$, $\check{T} = 0$; that is, the initial t-value t_0 becomes the new origin.) In other words, we translate (or time-shift) taxis by to units to make it $\check{T}$-axis. In this example, we will accordingly set $\check{T} = t - 10000$. The immediate effect is that it makes the initial conditions to be back at 0: $y(0) = 1$, $y'(0) = -7$, with respect to the new $\check{T}$-coordinate. We then solve the translated system of 2 equations to find C_1 and C_2. What we get is the (simpler) system,

$$1 = y(0) = C_1 e^0 + C_2 e^0 = C_1 + C_2$$
$$-7 = y'(0) = -C_1 e^0 - 4C_2 e^0 = -C_1 - 4C_2$$

the solution is $C_1 = -1$, and $C_2 = 2$. Hence, the solution, in the new $\check{T}$-coordinate system, is $y(\check{T}) = -e^{-\check{T}} + 2e^{-4\check{T}}$.

Lastly, since this solution is in terms of $\check{T}$, but the original problem was in terms of t, we should convert it back to the original context. This conversion is easily achieved using the translation formula used earlier, $\check{T} = t - t_0 = t - 10000$. By replacing every occurrence of $\check{T}$ by $t - 1000$ in the solution, we obtain the solution, in its proper independent variable t.

$$y(t) = -e^{-(t-10000)} + 2e^{-4(t-10000)}.$$

Example: Consider the solution y(t) of the initial value problem,

$$y'' - 2y' - 8y = 0, \quad y(0) = \alpha, \quad y'(0) = 2\pi.$$

Depending on the value of α, as $t \to \infty$, there are 3 possible behaviors of y(t). Explicitly determine the possible behaviors and the respective initial value α associated with each behavior.

The characteristic equation is $r^2 - 2r - 8 = (r+2)(r-4) = 0$. Its roots are $r = -2$ and 4. The general solution is then,

$$y = C_1 e^{-2t} + C_2 e^{4t}.$$

Notice that the long-term behavior of the solution is dependent on the coefficient C_2 only, since the C_1e^{-2t} term tends to 0 as $t \to \infty$, regardless of the value of C_1.

Solving for C_2 in terms of α, we get:

$$y(0) = \alpha = \quad C_1 + C_2$$
$$y'(0) = 2\pi = \ -2C_1 + 4C_2$$

$$2\alpha + 2\pi = 6C_2 \quad \to C_2 = \frac{\alpha + \pi}{3}.$$

Now, if $C_2 > 0$ then y tends to ∞ as $t \to \infty$. This would happen when $\alpha > -\pi$. If $C_2 = 0$ then y tends to 0 as $t \to \infty$. This would happen when $\alpha = -\pi$. Lastly, if $C_2 < 0$ then y tends to $-\infty$ as $t \to \infty$. This would happen when $\alpha < -\pi$.

- When $\alpha > -\pi$, $C_2 > 0$, $\lim_{t\to\infty} y(t) = \infty$.
- When $\alpha = -\pi$, $C_2 = 0$, $\lim_{t\to\infty} y(t) = 0$.
- When $\alpha < -\pi$, $C_2 < 0$, $\lim_{t\to\infty} y(t) = -\infty$.

The Existence and Uniqueness (of the Solution of a Second Order Linear Equation Initial Value Problem)

A sibling theorem of the first order linear equation Existence and Uniqueness Theorem.

Theorem: Consider the initial value problem,

$$y'' + p(t)y' + q(t)y = g(t), \quad y(t_0) = y_0, \quad y'(t_0) = y'_0.$$

If the functions p, q, and g are continuous on the interval I: $\alpha < t < \beta$ containing the point $t = t_0$. Then there exists a unique solution $y = \varphi(t)$ of the problem, and that this solution exists throughout the interval I.

That is, the theorem guarantees that the given initial value problem will always have (existence of) exactly one (uniqueness) twice-differentiable solution, on any interval containing t_0 as long as all three functions p(t), q(t), and g(t) are continuous on the same interval. Conversely, neither existence nor uniqueness of a solution is guaranteed at a discontinuity of p(t), q(t), or g(t).

Examples: For each IVP below, find the largest interval on which a unique solution is guaranteed to exist.

$$(t+2)y'' + ty' + \cot(t)y = t^2 + 1, \ y(2) = 11, \ y'(2) = -2.$$

The standard form is $y'' + \frac{t}{t+2}y' + \frac{\cos(t)}{(t+2)\sin(t)}y = \frac{t^2+1}{t+2}$, and $t_0 = 2$. The discontinuities of p, q, and g are $t = -2,\ 0,\ \pm\pi,\ \pm 2\pi,\ \pm 3\pi\ldots$ The largest interval that contains $t_0 = 2$ but none of the discontinuities is, therefore, $(0, \pi)$.

$$\sqrt{16 - t2}\ y'' + \text{In}(t+1)y' + \cos(t)y = 0,\quad y(0) = 2,\quad y'(0) = 0.$$

The standard form is $y'' + \dfrac{\text{In}(t+1)}{\sqrt{16-t^2}}y' + \dfrac{\cos(t)}{\sqrt{16-t^2}}y = 0,\ p(t)$

defined (and is continuous) on the interval (−1, 4), and similarly q(t) is only continuously defined on the interval (−4, 4); g(t) is continuous everywhere. Combining them we see that p, q, and g have discontinuities at any t such that $t \le -1$ or $t \ge 4$. That is, they are all continuous only on the interval (−1, 4). Since that interval contains $t_0 = 0$, it must be the largest interval on which the solution is guaranteed to exist uniquely. Therefore, the answer is (−1, 4).

Similar to the previous instance (first order linear equation version) of the Existence and Uniqueness Theorem, the only time that a unique solution is not guaranteed to exist anywhere is whenever the initial time t_0 occurs at a discontinuity of either p(t), q(t), or g(t).

Initial Value Problem vs. Boundary Value Problem

It might seem that there are more than one ways to present the initial conditions of a second order equation. Instead of locating both initial conditions $y(t_0) = y_0$ and $y'(t_0) = y'_0$ at the same point t_0, couldn't we take them at different points, for examples $y(t_0) = y_0$ and $y(t_1) = y_1$; or $y'(t_0) = y'_0$ and $y'(t_1) = y'_1$? The answer is NO. All the initial conditions in an initial value problem must be taken at the same point t_0. The sets of conditions above where the values are taken at different points are known as boundary conditions. A boundary value problem where a differential equation is bundled with (two or more) boundary conditions does not have the existence and uniqueness guarantee.

Example: Every function of the form y = Csin(t), where C is a real number satisfies the boundary value problem $y'' + y = 0$, $y(0) = 0$ and $y(\pi) = 0$. Therefore, the problem has infinitely many solutions, even though p(t) = 0, q(t) = 1, and g(t) = 0 are all continuous everywhere.

Fundamental Solutions

We have seen that the general solution of a second order homogeneous linear equation is in the form of $y = C_1y_1 + C_2y_2$, where y_1 and y_2 are two "distinct" functions both satisfying the given equation (as a result, y_1 and y_2 are themselves particular solutions of the equation). Now we will examine the circumstance under which two arbitrary solutions y_1 and y_2 could give us a general solution.

Suppose y_1 and y_2 are two solutions of some second order homogeneous linear equation such that their linear combinations $y = C_1y_1 + C_2y_2$ give a general solution of the equation.

Then, according to the Existence and Uniqueness Theorem, for any pair of initial conditions $y(t_0) = y_0$ and $y'(t_0) = y'_0$ there must exist uniquely a corresponding pair of coefficients C_1 and C_2 that satisfies the system of (algebraic) equations:

$$y_0 = C_1 y_1(t_0) + C_2 y_2(T_0)$$
$$y'_0 = C_1 y'_1(t_0) + C_2 y'_2(T_0)$$

From linear algebra, we know that for the above system to always have a unique solution (C_1, C_2) for any initial values y_0 and y'_0, the coefficient matrix of the system must be invertible, or, equivalently, the determinant of the coefficient matrix must be nonzero. That is,

$$\det\begin{pmatrix} y_1(t_0) & y_2(t_0) \\ y'(t_0) & y'_2(t_0) \end{pmatrix} = y1(t_0)y'_2(t_0) - y'_1(t_0)y_2(t_0) \neq 0$$

This determinant above is called the Wronskian or the Wronskian determinant. It is a function of t as well, denoted $W(y_1, y_2)(t)$, and is given by the expression,

$$W(y_1, y_2)(t) = y_1\, y'_2 - y'_1 y_2.$$

On the other hand, at each point t_0 where $W(y_1, y_2)(t_0) = 0$, a unique pair of coefficients C_1 and C_2 that satisfies the previous system of equations cannot always be found. This could be due to one of two reasons. The first reason is that $y = C_1\, y_1 + C_2\, y_2$ is really not a general solution of our equation. Or, the second possibility is that t_0 is a discontinuity of either p(t), q(t), or g(t). This second reason is, of course, a consequence of the Existence and Uniqueness theorem.

Assuming that not every point is a discontinuity of either p(t), q(t), or g(t), then the fact that $W(y_1, y_2)(t)$ is constant zero implies that $y = C_1\, y_1 + C_2\, y_2$ is not a general solution of the given equation. Otherwise, if $W(y_1, y_2)(t)$ is nonzero at some points to on the real line, then $y = C_1 y_1 + C_2\, y_2$ will, together with different combinations of initial condition $y(t_0) = y_0$ and $y'(t_0) = y'_0$, give uniquely all the possible particular solutions, on some open intervals containing t_0. That is, $y = C_1\, y_1 + C_2\, y_2$ is a general solution of the given equation. Hence, our interest in knowing whether or not $W(y_1, y_2)(t)$ is the constant zero function.

Formally, if $W(y_1, y_2)(t) \neq 0$, then the functions y_1, y_2 are said to be linearly independent. Else they are called linearly dependent if $W(y_1, y_2)(t) = 0$.

In the simple instance of two functions, as is the case presently, their linear independence could equivalently be determined by the fact that two functions are linearly independent if and only if they are not constant multiples of each other.

Suppose y_1 and y_2 are two linearly independent solutions of a second order homogeneous linear equation:

$$y'' + p(t)y' + q(t)y = 0.$$

That is, y_1 and y_2 both satisfy the equation, and $W(y_1, y_2)(t) \neq 0$. Then (and only then) their linear combination $y = C_1 y_1 + C_2 y_2$ forms a general solution of the differential equation. Therefore, a pair of such linearly independent solutions y_1 and y_2 is called a set of fundamental solutions, because they are essentially the basic building blocks of all particular solutions of the equation.

To summarize, suppose y_1 and y_2 are two solutions of a second order homogeneous linear equation, then:

$$W(y_1, y_2)(t) \text{ is not the constant zero function}$$

$$\updownarrow$$

$$y_1, y_2 \text{ are linearly independent}$$

$$\updownarrow$$

$$y_1, y_2 \text{ are fundamental solutions}$$

$$\updownarrow$$

$$y = C_1 y_1 + C_2 y_2 \text{ is a general solution of the equation}$$

Example: Let $y_1 = e^{r_1 t}$ and $y_2 = e^{r_2 t}$, $r_1 \neq r_2$ be any two different exponential function. Then,

$$W(y_1, y_2) = \det\begin{pmatrix} e^{r_1 t} & e^{r_2 t} \\ r_1 e^{r_1 t} & r_2 e^{r_2 t} \end{pmatrix} r_2 e^{r_1 t} e^{r_2 t} - r_1 e^{r_1 t} e^{r_2 t}$$

$$= (r_2 - r_1) e^{r_1 t} e^{r_2 t} \neq 0, \qquad \text{for all } t.$$

Therefore, any two different exponential-function solutions of a second order homogeneous linear equation (as those found using its characteristic equation) are always linearly independent, thus they will always give a general solution. Better yet, in this case since the Wronskian is never zero for all real numbers, a unique solution can always be found.

Lastly, here is an interesting (and, as we shall see shortly, useful) relationship between the Wronskian of any two solutions of a second order linear equation with its coefficient function p(t).

Abel's Theorem: If y_1 and y_2 are any two solutions of the equation,

$$y'' + p(t)y' + q(t)y = 0,$$

where p and q are continuous on an open interval I. Then the Wronskian $W(y_1, y_2)(t)$ is given by:

$$W(y_1, y_2)(t) = Ce^{-\int p(t)dt},$$

where C is a constant that depends on y_1 and y_2, but not on t. Further,m $W(y_1, y_2)(t)$ is either zero for all t in I (if $C = 0$) or else is never zero in I (if $C \neq 0$).

Case 2: Two Complex Conjugate Roots

When $b^2 - 4ac < 0$, the characteristic polynomial has two complex roots, which are conjugates $r_1 = \lambda + \mu i$ and $r_2 = \lambda - \mu i$ (λ, μ are real numbers, $\mu > 0$). As before they give two linearly independent solution $y_1 = e^{r_1 t}$ and $y_2 = e^{r_2 t}$. Consequently the linear combination $y = C_1 e^{r_1 t} + C_2 e^{r_2 t}$ will be a general solution. At this juncture you might have this question: "but aren't r_1 and r_2 complex numbers; what would become of the exponential function with a complex number exponent?" The answer to that question is given by the Euler's formula.

Euler's formula: For any real number θ,

$$e^{\theta i} = \cos\theta + i\sin\theta.$$

Hence, when r is a complex number $\lambda + \mu i$, the exponential function e^{rt} becomes:

$$e^{rt} = e^{(\lambda+\mu i)t} = e^{\lambda t} e^{\mu i t} = e^{\lambda t}(\cos \mu t + i \sin \mu t)$$

Similarly, when $r = \lambda - \mu i$, e^{rt} becomes:

$$e^{(\lambda-\mu i)t} = e^{\lambda t} e^{-\mu i t} = e^{\lambda t}(\cos(-\mu t) + i \sin(-\mu t))$$
$$= e^{\lambda t}(\cos \mu t + i \sin \mu t)$$

Hence, the general solution found above is the

$$y = C_1 e^{\lambda t}(\cos \mu t + i \sin \mu t) + C_2 e^{\lambda t}(\cos \mu t - i \sin \mu t)$$

However, this general solution is a complex-valued function (meaning that, given a real number t, the value of the function y(t) could be complex). It represents the general form of all particular solutions with either real or complex number coefficients. What we seek here, instead, is a real-valued expression that gives only the set of all particular solutions with real number coefficients only. In other words, we would like to "filter out" all functions containing coefficients with an imaginary part, that satisfy the given differential equation, keeping only those whose coefficients are real numbers.

$$u(t) = e^{\lambda t} \cos \mu t$$
$$v(t) = e^{\lambda t} \sin \mu t$$

It is easy to verify that both u and v satisfy the differential equation (one way to see this is to observe that u can be obtain from the complex-valued general solution by setting $C_1 = C_2 = 1/2$; and v can be obtained similarly by setting $C_1 = 1/2i$ and $C_2 = -1/2i$). Their Wronskian is $W(u, v) = \mu^{e2\lambda t}$ is never zero. Therefore, the functions u and v are linearly independent solutions of the equation. They form a pair of real-valued fundamental solutions and the linear combination is a desired real-valued general solution:

$$y = C_1 e^{\lambda t}\cos\mu t + C_2 e^{\lambda t}\sin\mu t.$$

When $r = \lambda \pm \mu i$, $\mu > 0$, are two complex roots of the characteristic polynomial.

Example: $y'' + 4y = 0$

$$y = C_1\cos 2t + C_2\sin 2t$$

$$y'' + 2y' + 5y = 0, \quad y(0) = 4, \quad y'(0) = 6$$

The characteristic equation is $r^2 + 2r + 5 = 0$, which has solutions $r = -1 \pm 2i$. So $\lambda = -1$ and $\mu = 2$. Therefore, the general solution is

$$y = C_1 e^{-t}\cos 2t + C_2 e^{-t}\sin 2t$$

Apply the initial conditions to find that $C_1 = 4$ and $C_2 = 5$. Hence,

$$y = 4e^{-t}\cos 2t + 5e^{-t}\sin 2t.$$

What would the solution be if the initial conditions are $y(25000) = 4$, and $y'(25000) = 6$ instead?

$$y = 4e^{-(t-25000)}\cos 2(t-25000) + 5e^{-(t-25000)}\sin 2(t-25000)$$

Case 3: One Repeated Real Root

When $b^2 - 4ac = 0$, the characteristic polynomial has a single repeated real root $r = \frac{-b}{2a}$. This causes a problem, because unlike the previous two cases the roots of characteristic polynomial presently only give us one distinct solution $y_1 = e^{rt}$. It is not enough to give us a general solution. We would need to come up with a second solution, linearly independent with y_1, on our own. How do we find a second solution?

Take what we have: a solution $y1 = e^{rt}$, where $r = \frac{-b}{2a}$. Let y_2 be another solution of the same equation $a y'' + b y' + c y = 0$. The standard form of this equation is $y'' + \frac{b}{a}y' + \frac{c}{a}y = 0$, where $p(t) = \frac{b}{a}$. Compute the Wronskian two different ways:

$$W(y_1, y_2) = \det\begin{pmatrix} e^{\lambda t} & y_2 \\ re^{rt} & y'_2 \end{pmatrix} = e^{\lambda t}y'_2 - re^{rt}\, y_2, \quad r = \frac{-b}{2a}$$

and

$$W(y_1, y_2) = Ce^{-\int p(t)dt} = Ce^{-\int \frac{b}{a}dt}, \quad C \neq 0.$$

By the Abel's Theorem, the fact C ≠ 0 guarantees that y_1 and y_2 are going to be linearly independent. Now, we have two expressions for the Wronskian of the same pair of solutions. The two expressions must be equal:

$$e^{\frac{-b}{2a}t} y'_2 + \frac{b}{2a} e^{\frac{-b}{2a}t} y_2 = Ce^{\frac{-b}{a}t}, \qquad C \neq 0.$$

This is a first order linear differential equation with y_2 as the unknown.

Put it into its standard form and solve by the integrating factor method.

$$y'_2 + \frac{b}{2a} y_2 = Ce^{\frac{-b}{2a}t}$$

The integrating factor is $\mu = e^{\int \frac{b}{2a}dt} = e^{\frac{b}{2a}t}$.

Hence,

$$y_2 = \frac{1}{e^{\frac{b}{2a}t}} \int e^{\frac{b}{2a}t} Ce^{\frac{-b}{2a}t} dt = e^{\frac{-b}{2a}t} \int C\,dt = e^{\frac{-b}{2a}t}(Ct + C_1)$$

$$= Cte^{\frac{-b}{2a}t} + C_1 e^{\frac{-b}{2a}t} = Cte^{rt} + C_1 e^{rt}$$

Any such a function would be a second, linearly independent solution of the differential equation. We just need one instance of such a function. The only condition for the coefficients in the above expression is C ≠ 0. Pick, say, C = 1, and C_1 = 0 would work nicely. Thus $y_2 = te^{rt}$.

Therefore, the general solution in the case of a repeated real root r is:

$$y = C_1 e^{rt} + C_2 te^{rt}.$$

Example: $y'' - 4y' + 4y = 0, \quad y(0) = 4, \quad y'(0) = 5$

The characteristic equation is $r^2 - 4r + 4 = (r-2)^2 = 0$, which has solution r = 2 (repeated). Thus, the general solution is:

$$y = C_1 e^{2t} + C_2 te^{2t}.$$

Differentiate,

$$y' = 2C_1 e^{2t} + C_2\left(2te^{2t} + e^{2t}\right).$$

Apply the initial conditions to find that $C_1 = 4$ and $C_2 = -3$:

$$y = 4e^{2t} - 3te^{2t}.$$

Reduction of Order

Given a second order, homogeneous, linear differential equation (with non-constant coefficients) and a known nonzero solution y_1, find the general solution of the given equation.

To start, assume that there exists a second solution in the form of $y_2 = y_1\, v(t)$, for some differentiable function v(t).

First we want to make sure the equation is written in the standard form with leading coefficient 1:

$$y''+p(t)y'+q(t)y=0.$$

Next, we will compute the Wronskian $W(y_1, y_2)(t)$ two different ways, using the two methods that we know. By the definition of Wronskian:

$$W(y_1, y_2)=\det\begin{pmatrix} y_1 & y_1 v(t) \\ y' & y' v(t)+y_1 v'(t)\end{pmatrix}=y_1 y_1' v(t)+y_1^2 v'(t)-y_1 y_1' v(t)=y_1^2 v'(t)$$

By the Abel's Theorem:

$$W(y_1, y_2)=Ce^{-\int p(t)dt}, \quad \text{where } C\neq 0.$$

The fact that $C \neq 0$ is important, because it guarantees the linear independence of y_1 and y_2.

The two expressions computed above are the Wronskian of the same two functions, therefore, the two expressions must be the same. Equate them:

$$y_1^2 v'(t)=Ce^{-\int p(t)dt}$$

Therefore, $v'(t)=Ce\dfrac{-\int p(t)dt}{y_1^2}, \qquad C\neq 0.$

Integrate the right-hand side to find v(t). Choose any convenient nonzero value for C. Letting C = 1 would work nicely, although it may not be the most convenient choice. Then find $y_2 = y_1\, v(t)$.

The general solution is still, of course, in the form $y = C_1 y_1 + C_2 y_2$. Therefore,

$$y=C_1 y_1+C_2 y_2=C_1 y_1+C_2 y_1 v(t).$$

It is actually not necessary to assume that $y_2 = y_1\, v(t)$. Although doing so makes the resulting first order differential equation easier to solve.

Example: If it is known that $y_1 = t$ is a solution of,

$$t^2y''-t(t+2)y'+(t+2)y=0, \qquad t>0.$$

Find its general solution.

Rewrite the equation into the standard form,

$$y''-\frac{t+2}{t}y'+\frac{t+2}{t^2}y=0$$

Identify $p(t)=-\frac{t+2}{t}$. Let $y_2=y_1v(t)=tv(t)$.

$$W(y_1,y_2)=\det\begin{pmatrix} t & tv(t) \\ 1 & v(t)+tv'(t)\end{pmatrix}tv(t)+t^2v'(t)-+tv(t)=t^2v'(t),$$

and $W(y_1,y_2)=Ce^{\int\frac{t+2}{t}dt}=Ce^{\int\left(1+\frac{2}{t}\right)}dt=Ce^{t+In\left(t^2\right)}=Ct^2e^t$,

where $C\neq 0$.

Equating both parts: $t^2v'=Ct^2e^t$

$$v'=Ce^t \quad \rightarrow \quad v=Ce^t+C_1$$

Choose $C = 1$ and $C_1 = 0 \rightarrow v = e^t$. Therefore, $y_2 = y_1$ v(t)= te^t.

The general solution is,

$$y=C_1y_1+C_2y_2=C_1t+C_2te^t.$$

Example: Find the general solution of the equation below, given that $y1=t^2\cos(\ln t)$ is a known solution.

$$t^2y''-3ty'+5y=0, \quad t>0.$$

Rewrite the equation into the standard form,

$$y''-\frac{3}{t}y'+\frac{5}{t^2}y=0$$

Identify $p(t)=-\frac{3}{t}$. Let $y_2=y_1v(t)=t_2\cos(\ln t)\ v(t)$.

$$W(y_1,y_2)=t^4\cos^2(\ln t)\ v'(t),$$

and

$$W(y_1,y_2)=Ce^{\int\frac{3}{t}dt}=Ce^{3In(t)}=Ce^{In\left(t^3\right)}=Ct^3, \quad C\neq 0.$$

Equating both parts: $t^4\cos^2(Int)v'=Ct^3$

$$v' = \frac{C}{t\cos^2(\text{Int})} = \frac{C\sec^2(\text{Int})}{t}$$

$$v = C\int \frac{\sec^2(\text{Int})}{t}dt = C\tan(\text{Int}) + C_1 = C\frac{\sin(\text{Int})}{\cos(\text{Int})} + C_1$$

Choose C = 1 and $C_1 = 0 \rightarrow v = \tan(\ln t)$.

$y_2 = y_1 v(t) = t^2 \cos(\ln t)\tan(\ln t) = t^2 \sin(\ln t)$.

The general solution is, $y = C_1 y_1 + C_2 y_2 = C_1 t^2 \cos(\ln t) + C_2 t^2 \sin(\ln t)$.

SYSTEMS OF LINEAR DIFFERENTIAL EQUATIONS

A system of linear differential equations is a set of linear equations relating a group of functions to their derivatives. Because they involve functions and their derivatives, each of these linear equations is itself a differential equation. For example, $f'(x) = f(x) + g(x)$ is a linear equation relating f' to f g, but $f' = fg$ is not, because the fg term is not linear. These equations can be solved by writing them in matrix form, and then working with them almost as if they were standard differential equations.

Systems of differential equations can be used to model a variety of physical systems, such as predator-prey interactions, but linear systems are the only systems that can be consistently solved explicitly.

System in Matrix Form

Given a system of linear differential equations of the form,

$$\begin{aligned} f_1'(x) &= a_{11}f_1(x) + \ldots + a_{1n}f_n(x), \\ f_2'(x) &= a_{21}f_1(x) + \ldots + a_{2n}f_n(x), \\ &\vdots \\ f_n'(x) &= a_{nl}f_1(x) + \ldots + a_{nn}f_n(x) \end{aligned}$$

this system is easiest to work with if it is rewritten as $v' = Av$ where:

$$v = \begin{bmatrix} f_1 \\ \vdots \\ f_n \end{bmatrix} \quad \text{and} \quad A = \begin{matrix} a_{11} & \cdots & a_{1n} \\ \vdots & \ddots & \vdots \\ a_{n1} & \cdots & a_{nn} \end{matrix}$$

Now, the n equations have been reduced to a single equation that is easier to work with.

Solving a System in Matrix Form

Because the matrix AA is constant, the equation $v'=Av$ resembles the equation $f'=cf$. The latter equation has solution $f=Ce^{cx}$, so it's tempting to guess that this equation has a solution involving an exponential as well. This guess turns out to be correct: the solution to this equation is $v(x)=e^{Ax}v(0)$, where we define:

$$e^{A}=\sum_{n-0}^{\infty}\frac{A^{n}}{n!}=I+A+\frac{A^{2}}{2}+\frac{A^{3}}{6}+\cdots,$$

using the power series for the exponential function. This fact can be verified by differentiating the power series.

Solve the system of differential equations,

$$f'(x)=g(x)\text{and}\,g'(x)=f(x),\ \text{where}\,f(0)=0\,\text{and}\,g(0)=1.$$

In matrix form, this system is $\begin{bmatrix} f \\ g \end{bmatrix}'=\underbrace{\begin{bmatrix} 0 & 1 \\ 1 & 0 \end{bmatrix}}_{A}\underbrace{\begin{bmatrix} f \\ g \end{bmatrix}}_{v}.$

If we compute some small powers of A, we find that $A^{2n}=I\,\text{and}\,A^{2n+1}=A$ i.e. the even powers are the identity and the odd powers are just A. Then,

$$e^{Ax}=\sum_{n-0}^{\infty}\frac{A^{n}}{n!}=\sum_{n-0}^{\infty}\frac{A^{2n}}{(2n!)}+\sum_{n-0}^{\infty}\frac{(Ax)^{2n+1}}{(2n+1)!}=A\sum_{n-0}^{\infty}\frac{x^{2n}}{(2n)!}+I\sum_{n-0}^{\infty}\frac{x^{2n+1}}{(2n+1)!}.$$

Now, we can recognize those two sums as the Taylor series for the hyperbolic sine and hyperbolic cosine, so we have,

$$e^{Ax}=A\sinh x+I\cosh x=\begin{bmatrix} \cosh x & \sinh x \\ \sinh x & \cosh x \end{bmatrix}.$$

Finally, we find

$$v(x)=e^{Ax}v(0)=\begin{bmatrix} \cosh x & \sinh x \\ \sinh x & \cosh x \end{bmatrix}\begin{bmatrix} 0 \\ 1 \end{bmatrix}=\begin{bmatrix} \sinh x \\ \cosh x \end{bmatrix},$$

so the solution is $f(x)=\sinh x$ and $g(x)=\cosh x$.

NONLINEAR DIFFERENTIAL EQUATIONS

A system of differential equations is said to be nonlinear if it is not a linear system. Problems involving nonlinear differential equations are extremely diverse, and methods of solution or analysis are problem dependent. Examples of nonlinear differential equations are the Navier–Stokes equations in fluid dynamics and the Lotka–Volterra equations in biology.

One of the greatest difficulties of nonlinear problems is that it is not generally possible to combine known solutions into new solutions. In linear problems, for example, a family of linearly independent solutions can be used to construct general solutions through the superposition principle. A good example of this is one-dimensional heat transport with Dirichlet boundary conditions, the solution of which can be written as a time-dependent linear combination of sinusoids of differing frequencies; this makes solutions very flexible. It is often possible to find several very specific solutions to nonlinear equations, however the lack of a superposition principle prevents the construction of new solutions.

MATRIX DIFFERENTIAL EQUATION

We consider matrix differential equations of the form:

$$M'(t)=AM(t)+U(t), \quad t\in\mathbb{C},$$

where A is a constant square matrix, U (t) is a given matrix function, and M(t) is an unknown matrix function. These equations appear often in many areas of mathematics and its applications, and finding their solutions is an important computational problem. The homogeneous case of $M'(t)=AM(t)+U(t), \quad t\in\mathbb{C}$, leads to the matrix exponential e^{tA}. This function is used to construct particular solutions of the non-homogeneous equations by means of the convolution product of functions.

We present some simple methods for the construction of solutions of $M'(t)=AM(t)+U(t), \quad t\in\mathbb{C}$,. We use the fact that finding a solution of the homogeneous matrix equation $M'(t)=AM(t)+U(t), \quad t\in\mathbb{C}$, reduces to finding a single solution g(t) of a scalar differential equation of the form w(D)g = 0, where w is any monic polynomial such that $w(A) = 0$. The solution e^{tA} corresponds to a solution g with certain initial values, called the dynamic solution. We present explicit constructions for the dynamic solution. Higher order matrix differential equations are also considered.

$$M(k+1)=AM(k)+U(k),\ k\geq 0,$$

which is a matrix difference equation, and leads to the sequence of powers A^k in the homogeneous case.

Construction of the Matrix Exponential

Let m be a positive integer. We denote by M the complex vector space of m × m matrixvalued functions of the complex variable t. Let M(t) be an element of M and let A be a constant m × m matrix. We consider first the homogeneous equation,

$$M'(t) = AM(t)$$

It is well-known that the solution of this equation that satisfies M(o) = I is the matrix exponential function e^{tA}, usually defined by the series,

$$e^{tA} = \sum_{k=o}^{\infty} \frac{t^k}{k!} A^k.$$

Let,

$$w(t) = b_0 t^{n+1} + b_1 t^n + b_2 t^{n-1} + \cdots + b_{n+1}$$

be any polynomial, with $b_0 = 1$, such that $w(A) = o$. It is clear that w may be any monic polynomial that is divisible by the minimal polynomial of A. By the Cayley–Hamilton theorem, $w(t) = \det(tI - A)$, the characteristic polynomial of A, is an acceptable w.

The equation $w(A) = o$ yields,

$$A^{n+1} = -\sum_{j=o}^{n} b_{j+1} A^{n-j}$$

and thus,

$$A^{n+1+k} = -\sum_{j=o}^{n} b_{j+1} A^{n+k-j}, \ k \geq o$$

Therefore all the powers A^k , for k n + 1, can be written as a linear combination of $I, A, A^2, \ldots, A^n$. This suggests that the series for M(t) could be written in the form,

$$M(t) = \sum_{k=o}^{n} fk(t) A^{n-k}.$$

Suppose that $M'(t) = AM(t)$ has a solution of this form. In order to find the coefficient functions fk(t) we use first a method based on Putzer's ideas, and another simple procedure based on operator identities.

Substitution of $M(t) = \sum_{k=o}^{n} fk(t) A^{n-k}.$ in $M'(t) = AM(t)$ gives

$$\sum_{k=o}^{n} f_k'(t) A^{n-k} = fo(t) A^{n+1} + \sum_{k=1}^{n} f_k(t) A^{n+1-k}$$

Combining this equation with $A^{n+1} = -\sum_{j=o}^{n} b_{j+1} A^{n-j}$ and collecting terms, we get:

$$\{f_n'(t) + b_{n+1} f_o(t)\} 1 + \sum_{k=o}^{n-1} \{f_k'(t) - f_{k+1}(t) + b_{k+1} f_o(t)\} A^{n-k} = o$$

For this equation to hold it is sufficient that the coefficients of the powers of A be zero. Note that the matrices I, A, A2,...,An need not be linearly independent. Then we look for functions f_k that satisfy the n + 1 equations:

$$f_{k+1}(t) - f_k'(t) = b_{k+1}f_o(t), \quad o \le k \le n-1,$$

and

$$-f_n'(t) = b_{n+1}f_o(t).$$

Differentiating the kth equation n – k times in $f_{k+1}(t) - f_k'(t) = b_{k+1}f_o(t), \ o \le k \le n-1,$ we obtain,

$$f_{k+1}^{(n-k)}(t) - f_k^{(n+1-k)}(t) = b_{k+1}f_o^{(n-k)}(t), \qquad o \le k \le n-1.$$

Adding all these equations with $-f_n'(t) = b_{n+1}f_o(t).$ we obtain a telescopic sum that yields,

$$\sum_{k=o}^{n+1} b_k D^{n+1-k} f_o(t) = o.$$

This equation can be written in the form $w(D)f_o(t) = o$, which is a scalar homogeneous linear differential equation with constant coefficients.

Equation $f_{k+1}(t) - f_k'(t) = b_{k+1}f_o(t), \ o \le k \le n-1,$ can be written as:

$$f_{k+1}(t) = f_k'(t) + b_{k+1}f_o(t), \quad o \le k \le n-1,$$

and by repeated substitution yields,

$$f_k(t) = (D^k + b_1D^{k-1} + b_2D^{k-2} + \cdots + b_k)f_o(t), \quad 1 \le k \le n,$$

that gives us $f_1, f_2, \ldots, f_n$ in terms of f_o. Therefore, finding a solution of $M'(t) = AM(t)$ of the form $M(t) = \sum_{k=o}^{n} f_k(t)A^{n-k}$. is reduced to finding a solution of the scalar equation $w(D)g(t) = o$.

From $M(t) = \sum_{k=o}^{n} f_k(t)A^{n-k}$. it is clear that we can have M(o)=I if $f_n(o)=1$ and $f_k(o)=o$ for $o \le k \le n-1$. In view of $f_k(t) = (D^k + b_1D^{k-1} + b_2D^{k-2} + \cdots + b_k)f_o(t), \ 1 \le k \le n,$ these initial values for the fk are obtained if f_o satisfies $D^k f_o(o) = o$ for $o \le k \le n-1$. and $D^n f_o(o) = 1$. The solution fo of $w(D)g(t) = o$ with such initial values I called the dynamic solution associated with w.

Given a square matrix A and a monic polynomial w such that $w(A)=o$, the solution M(t) of $M'(t)=AM(t)$ that satisfies M(o)=I is given by $M(t)=\sum_{k=0}^{n} f_k(t)A^{n-k}$, where the coefficients fk are given by:

$f_k(t)=(D^k+b_1D^{k-1}+b_2D^{k-2}+\dots+b_k)f_0(t),\quad 1\le k\le n$, the b_k are the coefficients of w, and f_0 is the dynamic solution associated with w.

That $f_k(t)=(D^k+b_1D^{k-1}+b_2D^{k-2}+\dots+b_k)f_0(t),\quad 1\le k\le n$ shows that $f_1, f_2,\dots,f_n$ are also solutions of $w(D)g = 0$ and, in view of $M(t)=\sum_{k=0}^{n} f_k(t)A^{n-k}$. All the entries of M(t) are also solutions of such equation. Therefore M(t) satisfies $w(D)M(t) = 0$.

Before we give a direct construction of the dynamic solution f_0, we present another approach to the construction of the matrix exponential. Note that,

$f_{k+1}(t)=f_k'(t)+b_{k+1}f_0(t),\quad 0\le k\le n-1$, is a sort of differential Horner's algorithm and that $f_k(t)=(D^k+b_1D^{k-1}+b_2D^{k-2}+\dots+b_k)f_0(t),\quad 1\le k\le n$ describes the f_k as the intermediate results in a sort of "synthetic division" computation that ends with $f_{n+1}(t) = w(D)f_0(t) = 0$. This is our motivation for the following development.

Let $w(t)$ be a monic polynomial of degree n + 1 with complex coefficients as in,

$w(t) = b_0t^{n+1}+b_1t^n+b_2t^{n-1}+\dots+b_{n+1}$. Define the difference quotient $w[t,x]$ by:

$$w[t,x]=\frac{w(t)-w(x)}{t-x}.$$

The basic polynomial identity,

$$t^{k+1}-x^{k+1}=(t-x)\sum_{j=0}^{k} t^j x^{k-j},\quad k\ge 0,$$

clearly shows that $w[t,x]$ is a symmetric polynomial in t and x of degree n in each variable.

Thus, $(t-x)w[t,x]=w(t)-w(x)$ is a polynomial identity.

Let D denote differentiation with respect to the complex variable t, and let A be an m × m matrix. Note that, on the elements of M, the operator D acts entrywise, A acts by multiplication on the left, and the operators D and A commute.

Replacing t by D and x by A in $w[t,x]$ we obtain a symmetric polynomial $w[D,A]$, which is a linear differential operator of order n with constant matrix coefficients. Since $(t-x)w[t,x]=w(t)-w(x)$ is a polynomial identity and D and A commute, w[D,A] satisfies the operator identity,

$$(D-A)w[D,A]=w(D)-w(A).$$

Suppose now that A satisfies $w(A) = 0$ and let f (t) be any scalar function such that w(D) f (t) = 0. Apply the operators that appear in $(D-A)w[D,A]=w(D)-w(A).$ to f (t)I , where I denotes the identity matrix, to obtain,

$$(D-A)w[A,D]f(t)I=w(D)f(t)I-w(A)f(t)I=0.$$

Therefore the matrix function M(t) = w[A,D]f (t)I is a solution of $M'(t)=AM(t)$. Note that M(t) is a polynomial in A of degree n with coefficients that depend on t. It is also a linear combination of the functions $D^k f$ (t), for $0\le k\le n$, with coefficients that are polynomials in A with degree less than or equal to n.

Using $t^{k+1}-x^{k+1}=(t-x)\sum_{j=0}^{k}t^j x^{k-j},\quad k\ge 0,$ we obtain,

$$w[t,x]=\sum_{k=0}^{n}bk\sum_{j=0}^{n-k}t^j x^{n-k-j}=\sum_{j=0}^{n}\left\{\sum_{K=0}^{n-j}b_k x^{n-k-j}\right\}t^j=\sum_{j=0}^{n}w_{n-j}\left(xt^j\right)$$

where the polynomials w_k are defined by:

$$w_k(x)=b_0x^k+b_1x^{k-1}+\cdots+bk,\ 0\le k\le n+1.$$

Note that,

$$w_{k+1}(x)=xw_k(x)+b_{k+1},\ 0\le k\le n,$$

and w_{n+1}(x) = w(x). We call the wk the Horner polynomials of w because above is Horner's algorithm for the computation of w(x). It is clear that $\{w_k\text{: } 0\le k\le n\ \}$ is a basis for the complex vector space of polynomials of degree less than or equal to n. This basis is also called the control basis associated with w.

Using $w[t,x]=\sum_{k=0}^{n}bk\sum_{j=0}^{n-k}t^j x^{n-k-j}=\sum_{j=0}^{n}\left\{\sum_{K=0}^{n-j}b_k x^{n-k-j}\right\}t^j=\sum_{j=0}^{n}w_{n-j}\left(xt^j\right)$ and the symmetry of w[t,x] we can write w[A,D] as follows,

$$w[A,D]=\sum_{j=0}^{N}A^{n-j}w_j=\sum_{j=0}^{n}w_{n-j}(A)D^J.$$

Therefore, for M(t) = w[A,D]f (t)I we have the expressions,

$$M(t)=\sum_{j=0}^{n}A^{n-j}w_j(D)f(t)=\sum_{j=0}^{n}w_{n-j}(A)D^j f(t).$$

The last expression shows that taking suitable values for $D^k f(0)$, for $0 \le k \le n$, we can obtain any polynomial in A as the value of M(0). In particular, if f (t) is the dynamic solution associated with w then M(0) = I , and hence M(t) = e^{tA}. We summarize the above results in the following.

Theorem: Let w be a monic polynomial and let $w_0, w_1, \ldots, w_n$ be the Horner polynomials of w. Let f (t) be the dynamic solution associated with w. Let A be any square matrix such that $w(A) = 0$. Then we have:

$$e^{tA} = \sum_{j=0}^{n} A^{n-j} w_j(D) f(t),$$

and

$$e^{tA} = \sum_{j=0}^{n} w_j(A) D^{n-j} f(t).$$

Note that the dependence on t of e^{tA} is completely determined by the dynamic solution f (t), which is in turn determined by the polynomial w. Note also that we can use the recurrence= relation $w_{k+1}(x) = x w_k(x) + b_{k+1}$, $0 \le k \le n$, to compute the summands in above.

The Dynamic Solution Associated with w

We present first a simple construction for the coefficients of the power series representation of the dynamic solution associated with w. An explicit expression for the dynamic solution as an exponential polynomial will be given later.

Let w be given by $w(t) = b_0 t^{n+1} + b_1 t^n + b_2 t^{n-1} + \cdots + b_{n+1}$. Define the reversed polynomial $w^*(t) = t^{n+1} w(1/t)$. Note that,

Let the formal power series $h(t) = \sum_{k \ge 0} h_k t^k$ be the reciprocal of $w^*(t)$. That is, h(t) $w^*(t) = 1$. Then the coefficients h_k satisfy $h_0 = 1$ and $h_k b_0 + h_{k-1} b_1 + \cdots + h_0 b_k = 0, \quad k \ge 1$.

Since $b_0 = 1$, this is a linear recurrence relation that can be used to compute all the h_k recursively. Using Cramer's rule on the system of linear equations that consists of $h_0 = 1$ and the equations obtained from $h_k b_0 + h_{k-1} b_1 + \cdots + h_0 b_k = 0, \quad k \ge 1$. by taking $1 \le k \le j$, we obtain,

$$h_j = (-1)^j \det \begin{bmatrix} b_1 & 1 & & & \\ b_2 & b_1 & 1 & & \\ \vdots & \vdots & \vdots & \ddots & \\ b_{j-1} & b_{j-2} & b_{j-3} & \cdots & 1 \\ b_j & b_{j-1} & b_{j-2} & \cdots & b_1 \end{bmatrix}, \quad j \ge 1.$$

Recall that $b_j = 0$ if $j>n + 1$ and thus the matrix above is banded if $j>n + 1$. This determinant representation for the h_j shows the complexity of the dependence of h_j on the coefficients $b_1, b_2,...,b_j$, but it is not a practical way to compute the h_j.

$$f_w(t)=\sum_{k=0}^{\infty} hk\frac{t^{k+n}}{(k+n)!}.$$

We claim that $f_w(t)$ is the dynamic solution associated with w. The convergence of above for all complex values of t is a consequence of another expression for $f_w(t)$ that we will give later. An alternative proof is obtained by using the fact that the h_k are the coefficients of the Taylor series of the reciprocal of a polynomial, and that such series converge outside some disk.

Let us show first that $w(D)f_w = 0$. Since $D^j(t^k/k!) = t^{k-j}/(k-j)!$ for $j \le k$, and $D^j(t^k/k!) = 0$ for $j>k$, by a simple rearrangement of terms it is easy to see that,

$$\begin{aligned}
w(D)f_w(t) &= \sum_{k\ge 0}\sum_{j=0}^{n+1} h_k b_{n+1-j}\frac{t^{n+k-j}}{(n+k-j)!} \\
&= \sum_{k\ge 0}\sum_{l=k-1}^{n+k} h_k b_{l+1-k}\frac{t^l}{l!} \\
&= \sum_{l\ge 0}\sum_{k=l-n}^{l+1} h_k b_{l+1-k}\frac{t^l}{l!} = 0
\end{aligned}$$

The last equality follows from $h_k b_0 + h_{k-1} b_1 + \cdots + h_0 b_k = 0, \quad k \ge 1$. which shows that all the coefficients in the last power series are zero. Note that in the above sums we take $h_k = 0$ whenever $k < 0$.

Since t^n is a common factor of all the terms in the series for $f_w(t)$, and the coefficient of the first term is $h_0 = 1$, it is clear that $f_w(t)$ has the correct initial values to be the dynamic solution associated with w.

Any Taylor polynomial approximation of $f_w(t)$ obtained by truncation of $f_w(t)=\sum_{k=0}^{\infty} hk\frac{t^{k+n}}{(k+n)!}$. yields immediately a polynomial approximation for e^{tA}, by substitution of f_w by its approximating polynomial in $A^{n+1+k} = -\sum_{j=0}^{n} b_{j+1}A^{n+k-j}, \; k \ge 0$. We find next another representation for the dynamic solution. Let:

$$w(t)=\prod_{j=0}^{r}(t-a_j)^{mj+1},$$

where the a_j are distinct complex numbers and the m_j are nonnegative integers such that $\sum_j (m_j+1) = n+1$. We consider first the differential operators of the form $(D - aI)^k$, because w(D) is a product of a finite number of such operators. The basic exponential polynomials are the functions,

$$g_{a,k}(t)=\frac{t^k}{k!}e^{at},\ a\in\mathbb{C},\ k\in\mathbb{N},$$

defined on the complex plane. It is easy to show that these functions are linearly independent. Denote by E the complex vector space generated by all the $g_{a,k}$. Its elements are called exponential polynomials.

A simple computation gives:

$$(D-aI)^j g_{a,k}(t) = \begin{cases} g_{a,k-j}(t), & k\geq j \\ 0, & k<j. \end{cases}$$

In particular, any linear combination of the $g_{a,k}(t)$, for $0\leq k<j$ is an element of the null space of $(D-aI)^j$. Therefore, the vector space generated by the set $S=\{g_{aj,k}: 0\leq j\leq r,\ 0\leq k\leq m_j\}$ is contained in the null space of $w(D)$. In fact, it is easy to see that S is a basis for the null space of $w(D)$.

We define a commutative multiplication $*$ on E, called the convolution product, as follows. For $a\neq b$

$$g_{a,m} * g_{b,k} \sum_{j=0}^{m} C(a,\ j,\ b,\ k) g_{a,m-j} + \sum_{j=0}^{k} C(b,\ j,\ a,\ m) g_{b,k-j},$$

where the coefficients are defined by:

$$C(a,\ j,\ b,\ i)=(-1)^j\binom{J+i}{j}(a-b)^{-j-i-1},\ a\neq b,\ i,\ j\in\mathbb{N}$$

and for the remaining case define:

$$g_{a,k} * g_{a,m} = g_{a,1+k+m},\ a\in\mathbb{C},\ k,m\in\mathbb{N}.$$

Let Φ be the linear functional on ε defined by $\Phi g_{a,k}=\delta_{0,k}$, for (a, k) in $\mathbb{C}\times\mathbb{N}$. Note that Φ can be seen as evaluation at t = 0. A straightforward computation gives:

$$D(g_{a,k} * g_{b,m}) = (D_{ga,k}) * g_{b,m} + g_{b,m}\Phi g_{a,k},$$

and by linearity we have:

$$D(g*f) = (Dg)*f + f\Phi g,\ f,\ g\in\varepsilon.$$

Theorem: Let w(t) be as in $\omega(t)=\prod_{j=0}^{r}(t-a_j)^{mj+1}$, and let

$$f_w = g_{a0,m0} * g_{a1,m1} *\cdots* g_{ar,mr}.$$

Then f_w is the dynamic solution associated with w,

Proof: From the definition of the convolution it is clear that the subspace of ε generated by S is closed under the convolution product and thus f_w is in the null space of $w(D)$.

From $g_{a,m} * g_{b,k} \sum_{j=0}^{m} C(a, j, b, k) g_{a,m-j} + \sum_{j=0}^{k} C(b, j, a, m) g_{b,k-j}$

and,

$$C(a, j, b, i) = (-1)^j \binom{J+i}{j} (a - b)^{-j-i-1}, \ a \neq b, \ i, j \in \mathbb{N} \text{ we get,}$$

$$\Phi(g_{a,m} * g_{b,k}) = C(a, m, b, k) + C(b, k, a, m) = 0$$

for $a \neq b$ and any m and k in N, and from $g_{a,k} * g_{a,m} = g_{a,1+k+m}$, $a \in \mathbb{C}$, $k, m \in \mathbb{N}$.

$$\Phi(g_{a,k} * g_{a,m}) = \Phi g_{a,k+m+1} = 0, \ k, m \in \mathbb{N}.$$

Therefore the value at zero of the convolution product of any two basic exponential polynomials is zero. Thus we have,

$$\Phi(g_{a,k} * f) = 0, \ (a,k) \in \mathbb{C} \times \mathbb{N}, \ f \in \varepsilon.$$

We determine next the initial values of f_w. Consider first the case when w has only one root a of multiplicity $n + 1$ and let $0 \leq k \leq n$. Then $f_w = g_{a,n}$ and writing $D^k = ((D - aI) + aI)^k$ we get:

$$D^k f_w = \sum_{\ell=0}^{k} \binom{k}{\ell} a^{k-\ell} (D - aI)^\ell g_{a,n} = \sum_{\ell=0}^{k} \binom{k}{\ell} a^{k-\ell} g_{a,n-\ell}$$

and thus,

$$\Phi D^k f_w = \sum_{\ell=0}^{k} \binom{k}{\ell} a^{k-\ell} \delta_{\ell,n} = \delta_{k,n}.$$

Suppose now that w has at least two distinct roots, let n > 1 and let $0 \leq k \leq n$.. Using $g_{a,k} * g_{a,m} = g_{a,1+k+m}$, $a \in \mathbb{C}$, $k, m \in \mathbb{N}$, we can write fw = g ∗ h where g = $g_{ao,o}$ and h is the convolution of the $g_{ai,mi}$ for i > 0 and also $g_{ao,mo-1}$ if $m_o > 0$. Note that h = f_u where u(t) = w(t)/(t − a_o) and u(t) has degree equal to n. This fact is used in the induction step below. Note also that $\Phi D^j g = a_o^j$ for $j \geq 0$. Then $D(g * h) = Dg * h + h\Phi g$ and by iteration we obtain,

$$D^k (g * h) = D^k g * h + \sum_{\ell=0}^{k-1} a_o^{k-1-\ell} D^\ell h.$$

Applying Φ to both sides of this equation and using $\Phi(g_{a,k} * f) = 0,\ (a,k) \in \mathbb{C} \times \mathbb{N},\ f \in \varepsilon.$ and induction on n we get:

$$\Phi D^k(g * h) = \sum_{\ell=0}^{k-1} a_0^{k-1-\ell} D^\ell \delta_{\ell,n-1} = \delta_{k-1,n-1} = \delta_{k,n}.$$

Therefore $f_w = g * h$ has the correct initial values to be the dynamic solution associated with w.

By the uniqueness theorem for the solutions of linear differential equations with constant coefficients we see that the exponential polynomial f_w defined above must coincide with the power series f_w defined in $f_w(t) = \sum_{k=0}^{\infty} hk \frac{t^{k+n}}{(k+n)!}$.

The computation of f_w using $f_w = g_{a0,m0} * g_{a1,m1} * \cdots * g_{ar,mr}$ is a straightforward repeated application of the convolution formula $g_{a,m} * g_{b,k} \sum_{j=0}^{m} C(a, j, b, k) g_{a,m-j} + \sum_{j=0}^{k} C(b, j, a, m) g_{b,k-j}$,

It is in fact essentially equivalent to finding the partial fractions decomposition of $1/w(t)$.

A simple computation gives us:

$$(D - aI)^{k+1}(g_{a,k} * h) = h, (a,k) \in \mathbb{C} \times \mathbb{N}, h \in \varepsilon.$$

This means that the map that sends h to $g_{a,k} * h$ is a right inverse for $(D - aI)^{k+1}$. Applying this result repeatedly we obtain the following.

Theorem: Let w(t) be a polynomial as in $w(t) = \prod_{j=0}^{r} (t - a_j)^{mj+1}$, and let fw be given by $f_w = g_{a0,m0} * g_{a1,m1} * \cdots * g_{ar,mr}$. Then the linear map on E that sends h to $f_w * h$ is a right inverse for the operator w(D).

The ideas used in the proof of previous Theorem can be used to show that the particular solution $f_w * h$ of $w(D)g = h$ has all its initial values equal to zero.

Non-homogeneous Matrix Equations

Let us consider now the non-homogeneous matrix equation,

$$M'(t) = AM(t) + U(t),$$

where A is a constant m × m matrix, U (t) is a given element of $\mathcal{M}$ and M(t) is an unknown matrix function. Let,

$$U(t) = \sum_{j=0}^{s} \alpha_j(t) B_{s-j},$$

where the B_j are constant m × m matrices and suppose that the $\alpha_j(t)$ are exponential polynomials.

Let $w(t)=t^{n+1}+b_1t^n+\dots+b_{n+1}$ be a monic polynomial such that w(A) = o, let $f_o(t)$ be the dynamic solution associated with w, and let $f_k(t)$ be defined by, $f_k(t)=(D^k+b_1D^{k-1}+b_2D^{k-2}+\dots+b_k)f_o(t),\quad 1\le k\le n,$. Then we have,

$$e^{tA}=\sum_{k=o}^{n}f_k(t)A^{n-k}.$$

We define the convolution,

$$e^{tA}*U(t)=\sum_{k=o}^{n}\sum_{j=o}^{s}\left(f_k(t)*\alpha_j(t)\right)A^{n-k}B_{s-j}.$$

Note that this convolution is not commutative because of the noncommutativity of the matrix multiplication.

Theorem: The matrix function $e^{tA}*U(t)$ is a particular solution of $M'(t)=AM(t)+U(t)$,

Proof: By $D(g*f)=(Dg)*f+f\Phi g$, f, $g\in\varepsilon$. we have,

$$D(f_k(t)*\alpha_j(t))=f_k'(t)*\alpha_j(t)+\alpha_j(t)f_k(o)$$

and $f_k(o)=\delta_{k,n}$ by $f_k(t)=(D^k+b_1D^{k-1}+b_2D^{k-2}+\dots+b_k)f_o(t),\quad 1\le k\le n$, because f_o is the dynamic solution. Therefore,

$$D\left(e^{tA}*U(t)\right)=\sum_{k=o}^{n}\sum_{j=o}^{s}\left(f_k'(t)*\alpha_j(t)\right)A^{n-k}B_{s-j}\sum_{k=o}^{n}\sum_{j=o}^{s}\delta_{k,n}\alpha_j(t)A^{n-k}B_{s-j}.$$

The last sum is clearly equal to U (t). Since e^{tA} is a solution of M(t)=AM(t) the first sum equals $Ae^{tA}*U(t)$. This completes the proof.

An apparent limitation of the previous theorem is the hypothesis that the functions $\alpha_j(t)$ must be exponential polynomials. We can eliminate this limitation as follows. Note that the proof of the theorem only uses the fact that the convolutions $f_k(t)*\alpha_j(t)$ are well defined and that $D(f_k(t)*\alpha_j(t))=f_k'(t)*\alpha_j(t)+\alpha_j(t)f_k(o)$ holds.

The Duhamel convolution of functions f and g of a real or complex variable t is defined by:

$$f*g(t)=\int_o^t f(t-z)g(z)dz.$$

Using elementary tools it is easy, but quite laborious, to show that the convolution defined in $g_{a,m} * g_{b,k} \sum_{j=0}^{m} C(a, j, b, k) g_{a,m-j} + \sum_{j=0}^{k} C(b, j, a, m) g_{b,k-j}$, and the Duhamel convolution coincide on the space E of exponential polynomials. We can allow the $\alpha_j(t)$ to be any functions for which their Duhamel convolution with any basic exponential polynomial is well defined. For example, they can be piecewise continuous real functions. Using basic properties of integration it is easy to show that $D(f_k(t) * \alpha_j(t)) = f_k'(t) * \alpha_j(t) + \alpha_j(t) f_k(0)$ also holds in this case.

Higher Order Equations

We consider now linear matrix differential equations of order r with matrix coefficients. Let $B_0, B_1, \ldots, B_r$ be m × m matrices, with $B_0 = I$. Define the differential operator,

$$L = \sum_{j=0}^{r} B_j D^{r-j},$$

where D denotes differentiation with respect to t. We want to find a matrix function F (t) such that,

$$LF(t) = \sum_{j=0}^{r} B_j D^{r-j}(t) = 0,$$

for $t \in \mathbb{C}$, and

$$F^{(j)}(0) = Q_j, \quad 0 \le j \le r-1,$$

where the Q_j are given matrices.

Define the block companion matrix,

$$A = \begin{bmatrix} 0 & I & 0 & \cdots & 0 & 0 \\ 0 & 0 & I & \cdots & 0 & 0 \\ \vdots & \vdots & \vdots & \cdots & \vdots & \vdots \\ 0 & 0 & 0 & \cdots & 0 & I \\ -Br & -B_{r-1} & -B_{r-2} & \cdots & -B_2 & -B_1 \end{bmatrix},$$

where the entries are m × m matrices. Let $M(t) = e^{tA}$. Then we have M(t) = AM(t) and $M(0) = I$. We consider M(t) as an r × r matrix whose entries $M_{i,j}(t)$ are m × m matrices. Since M (t) = AM(t), it is easy to see that,

$$M_{i,k}(t) = M_{1,k}^{(i=1)}(t), \quad 1 \le i \le r, \; 1 \le k \le r.$$

and

$$LM_{1,k}(t) = 0, \; 1 \le k \le r.$$

Since M(t) commutes with A we have M (t) = M(t)A, and looking at the first row in this matrix equation we get:

$$M'_{1,k}(t) = M_{1,k-1}(t) - M_{1,r}(t)B_{r+1-k},\ 1 \le k \le r,$$

where $M_{1,0}(t) = 0$. Solving for $M_{1,k-1}(t)$ and taking k = r, r – 1,...,r – j + 1 we obtain,

$$\begin{aligned} M_{1,r-1}(t) &= M'_{1,r}(t) + M_{1,r}(t)B_1, \\ M_{1,r-2}(t) &= M'_{1,r-1}(t) + M_{1,r}(t)B_2, \\ &\vdots \\ M_{1,r-j}(t) &= M'_{1,r-j+1}(t) + M_{1,r}(t)B_j. \end{aligned}$$

By repeated differentiation and forward substitution we obtain,

$$M_{1,r-j}(t) = M_{1,r}^{(j)}(t) + M_{1,r}^{(j-1)}(t)B_1 + \cdots + M_{1,r}(t)B_j,\ 1 \le j \le r-1.$$

We denote by $L_jM_{1,r}(t)$ the expression in the right-hand side.

$$F(t) = \sum_{k=1}^{r} Q_{k-1}M_{1,k}(t)\sum_{k=1}^{r} Q_{r-k}L_{k-1}M_{1,r}(t).$$

From $LM_{1,k}(t) = 0,\ 1 \le k \le r.$ we see that F (t) satisfies $LF(t) = \sum_{j=0}^{r} B_jD^{r-j}(t) = 0,$. Since M(0) = I , by $M_{i,k}(t) = M_{1,k}^{(i=1)}(t),\ \ 1 \le i \le r,\ 1 \le k \le r.$ we have,

$$M_{i,k}(0) = \delta_{i,k}I = M_{1,k}^{(i-1)}(0),\ 1 \le i \le r,\ 1 \le k \le r.$$

Therefore F (t) satisfies $F^{(j)}(0) = Q_j,\ 0 \le j \le r-1,$. Note that F (t) is expressed in terms of the matrix function $M_{1,r}(t)$. This is a solution of $LF(t) = \sum_{j=0}^{r} B_jD^{r-j}(t) = 0,$ with initial values:

$$M_{1,r}^{(r-1)}(0) = I,\quad M_{1,r}^{(j)}(0) = 0,\quad 0 \le j \le r-2.$$

We call $M_{1,r}(t)$ the dynamic solution of $LF(t) = \sum_{j=0}^{r} B_jD^{r-j}(t) = 0,$

Since $M_{1,r}(t)$ is a block of e^{tA},

Let w be the characteristic polynomial of A. Let n = rm – 1 and f be the dynamic solution associated with w. Then, by $e^{tA} = \sum_{j=0}^{n} A^{n-j}w_j(D)f(t),$ we have:

$$M(t) = \sum_{j=0}^{n} w_{n-j}(D)f(t)A^j.$$

Since A is a block companion matrix, if we write the last column of A_j in the form $\left[P_j P_{j+1} \cdots P_{j+r-1}\right]^T$, for $j \geq 0$, then the sequence P_j of m × m matrices satisfies the linear recurrence relation,

$$P_{j+r} = -\sum_{\ell=1}^{r} B_\ell P_{j+r-\ell}, \quad j \geq 0,$$

and has initial values $P_j = 0$ for $0 \leq j \leq r-2$, and $P_{r-1} = I$. The sequence P_j is called the dynamic solution of above.

Therefore, the block in the (1,r) position of M(t) is,

$$M_{1,r-j}(t) = \sum_{j=0}^{n} w_{n-j}(D) f(t) P_j$$

Note that $M_{1,r}(t)$ is completely determined by the characteristic polynomial of A and the first rm terms of the sequence P_j, which appear in the last column of the block decomposition of I,A,A²,...,An.

Combining above and $F(t) = \sum_{k=1}^{r} Q_{k-1} M_{1,k}(t) \sum_{k=1}^{r} Q_{r-k} L_{k-1} M_{1,r}(t)$. we obtain,

$$F(t) = \sum_{k=1}^{r} Q_{r-k} L_{k-1} \left(\omega_{n-j}(D) f(t) P_j\right).$$

Matrix Difference Equations

We present here some examples of matrix difference equations.

For any scalar or matrix-valued function $f(\ell)$, defined for in the set $\mathbb{N}$ of nonnegative integers, the forward shift operator E is defined by $Ef(\ell) = f(\ell+1)$. Let A be a given m × m matrix. Consider the matrix difference equation,

$$M(\ell+1) = AM(\ell), \ \ell \geq 0.$$

It is obvious that the solution of this equation is $M(\ell) = A^\ell M(0)$.

From the polynomial identity $(t-x)w[t,x] = w(t) - w(x)$ we get the operator identity,

$$(E-A)w[E,A] = w(E) - w(A).$$

Suppose now that $w(A) = 0$ and let $g(\ell)$ be any scalar solution of the difference equation $w(E)f(\ell) = 0$. Then, the operator identity above shows that the matrix function $M(\ell) = w[E,A]g(\ell)I$ is a solution of above. It is easy to see that if g is the solution with initial values $g(\ell) = 0$ for $0 \leq \ell \leq n-1$ and $g(n) = 1$, then $M(0) = I$ and $M(\ell) = A^\ell$ for all $\ell \geq 0$. Such sequence g is called the dynamic solution of $w(E)f(\ell) = 0$.

Theorem: Let w be a monic polynomial and let $w_0, w_1, \ldots, w_n$ be the Horner polynomials of w. Let $g(\ell)$ be the dynamic solution of the scalar difference equation $w(E)g = 0$. Let A be any square matrix such that $w(A) = 0$. Then we have,

$$A^{\ell} = \sum_{k=0}^{n} A^{n-k} w_k(E) g(\ell), \quad \ell \geq 0,$$

and

$$A^{\ell} = \sum_{k=0}^{n} g(\ell + n - k) w_k(A), \quad \ell \geq 0.$$

We find next a simple explicit construction of the dynamic solution $g(\ell)$. Define the basic linearly recurrent sequences,

$$s_{a,k}(\ell) = \binom{\ell}{k} a^{\ell-k}, \quad a \in \mathbb{C}, k \in \mathbb{N}.$$

Let S denote the complex vector space generated by the sequences $s_{a,k}(\ell)$. The elements of S are called linearly recurrent sequences.

Note that,

$$(E - aI)^j s_{a,k}(\ell) = \begin{cases} sa,k-j(\ell), & k \geq j \\ 0 & k < j. \end{cases}$$

In particular, any linear combination of the sequences $s_{a,k}(\ell)$ for $0 \leq k < j$ is an element of the null space of $(E - aI)^j$ Therefore, if $w(t)$ is as in $w(t) = \prod_{j=0}^{r} (t - a_j)^{mj+1}$, then the vector space generated by $\{s_{aj,k} : 0 \leq j \leq r, \ 0 \leq k \ m_j\}$ is contained in the null space of $w(E)$.

We define the convolution product $*$ on S as follows. For $a \neq b$

$$s_{a,m} * s_{b,k} = \sum_{j=0}^{m} C(a, j, b, k) s_{a,m-j} + \sum_{j=0}^{k} C(b, j, a, m) s_{b,k-j},$$

where the coefficients C(a, j, b, i) are defined in,

$$C(a, j, b, i) = (-1)^j \binom{J+i}{j} (a - b)^{-j-i-1}, \ a \neq b, \ i, j \in \mathbb{N}, \text{ and } s_{a,m} * s_{a,k} = s_{a,1+m+k}.$$

Notice that the previous statements are obtained by putting E in the place of D, $s_{a,k}$ in the place of $g_{a,k}$, and in the place of t. We are using here a linear isomorphism between the spaces E and S determined by the correspondence $g_{a,k} \rightarrow s_{a,k}$. The operators E and D are similar in the linear algebraic sense. In a linear algebra approach is used to study functional equations determined by an operator L that is similar to the differentiation operator D.

The results about D and the convolution on E are transformed immediately to results about E and the convolution on S. For example, from theorem we obtain the following.

Theorem: Let $w(t)$ be as in $w(t)=\prod_{j=0}^{r}(t-a_j)^{mj+1}$, and let,

$$f_w = s_{a0,m0} * s_{a1,m1} *\cdots* s_{ar,mr}.$$

Then $f_w(\ell)$ is the dynamic solution associated with w.

In the same way, we can imitate the results to find particular solutions of the non-homogeneous difference equation,

$$M(\ell+1)=AM(\ell)+U(\ell),$$

where $U(\ell)$ is a given matrix sequence. If,

$$U(\ell)=\sum_{j=0}^{s}\alpha_j(\ell)B_{s-j}$$

where the B_j are constant m×m matrices and the $\alpha_j(\ell)$ are arbitrary sequences, and A is given by $A^{\ell}=\sum_{k=0}^{n}g(\ell+n-k)w_k(A), \quad \ell \geq 0$ then,

$$A^{\ell} * U(\ell)=\sum_{k=0}^{n}\sum_{j=0}^{s}\left(g(\ell+n-k)*\alpha_j(\ell)\right)wk(A)B_{s-j}$$

is a particular solution of $M(\ell+1)=AM(\ell)+U(\ell)$,

The extension to the case of higher order difference equations can be done using the method presented.

References

- Gintautas, V. (2008). "Resonant forcing of nonlinear systems of differential equations". Chaos. 18 (3): 033118. arXiv:0803.2252. Bibcode:2008Chaos..18c3118G. doi:10.1063/1.2964200. PMID 19045456
- Linear-differential-equations, differential-equations, maths, guides: toppr.com, Retrieved 15 March, 2020
- De Canete, Javier, Cipriano Galindo, and Inmaculada Garcia-Moral (2011). System Engineering and Automation: An Interactive Educational Approach. Berlin: Springer. p. 46. ISBN 978-3642202292. Retrieved 20 January 2018
- Systems-of-linear-ode: brilliant.org, Retrieved 16 May, 2020
- "Nonlinear Biology", The Nonlinear Universe, The Frontiers Collection, Springer Berlin Heidelberg, 2007, pp. 181–276, doi:10.1007/978-3-540-34153-6_7, ISBN 9783540341529

4

Partial Differential Equations: An Integrated Study

Partial differentiation equations comprise of an unknown multivariable and their partial derivatives. First order partial differential equations, second order partial differential equations, fundamental solution, nonlinear partial differential equation, etc. fall within it. All the aspects related to partial differential equations have been carefully analyzed in this chapter.

A differential equation involving partial derivatives of a dependent variable(one or more) with more than one independent variable is called a partial differential equation, hereafter denoted as PDE.

Consider the Following Equations:

$$\frac{\partial^2 u}{\partial t^2} - c^2 \frac{\partial^2 u}{\partial x^2} = 0$$

$$\frac{\partial u}{\partial t} - T \frac{\partial^2 u}{\partial x^2} = 0$$

$$\frac{\partial^2 u}{\partial x^2} + \frac{\partial^2 u}{\partial y^2} = 0$$

$$\left(x^2 + y^2\right) \frac{\partial u}{\partial t} + \frac{\partial^2 u}{\partial x \partial y} - 3u = 0$$

$$ux \frac{\partial^2 u}{\partial x^2} + u^2 xy \frac{\partial^2 u}{\partial x \partial y} + uy \frac{\partial^2 u}{\partial y^2} + \left(\frac{\partial u}{\partial x}\right)^2 + \left(\frac{\partial u}{\partial x}\right)^2 + u^3 = 0$$

$$\frac{\partial^2 u}{\partial x^2} + \left(\frac{\partial^2 u}{\partial x \partial y}\right)^2 + \frac{\partial^2 u}{\partial y^2} = x^2 + y^2$$

- Order of a PDE: The order of the highest derivative term in the equation is called the order of the PDE. Thus equations are all of second order.

- Linear PDE: If the dependent variable and all its partial derivatives occure linearly in any PDE then such an equation is called linear PDE otherwise a non-linear PDE.
- Quasi-linear PDE: A PDE is called as a quasi-linear if all the terms with highest order derivatives of dependent variables occur linearly, that is the coefficients of such terms are functions of only lower order derivatives of the dependent variables. However, terms with lower order derivatives can occur in any manner.
- Homogeneous PDE: If all the terms of a PDE contains the dependent variable or its partial derivatives then such a PDE is called non-homogeneous partial differential equation or homogeneous otherwise. In the above six examples in equation $\frac{\partial^2 u}{\partial x^2}+\left(\frac{\partial^2 u}{\partial x \partial y}\right)^2+\frac{\partial^2 u}{\partial y^2}=x^2+y^2$ is non-homogeneous whereas the first five equations are homogeneous.
- Notation: It is also a common practice to use subscript notation in writing partial differential equations. For example the Laplace Equation in three dimensional space:

  $$\frac{\partial^2 u}{\partial x^2}+\frac{\partial^2 u}{\partial y^2}+\frac{\partial^2 u}{\partial z^2}=0$$

  $$u_{xx}+u_{yy}+u_{xx}=0$$

Existence and Uniqueness

Although the issue of existence and uniqueness of solutions of ordinary differential equations has a very satisfactory answer with the Picard–Lindelöf theorem, that is far from the case for partial differential equations. The Cauchy–Kowalevski theorem states that the Cauchy problem for any partial differential equation whose coefficients are analytic in the unknown function and its derivatives, has a locally unique analytic solution. Although this result might appear to settle the existence and uniqueness of solutions, there are examples of linear partial differential equations whose coefficients have derivatives of all orders (which are nevertheless not analytic) but which have no solutions at all. Even if the solution of a partial differential equation exists and is unique, it may nevertheless have undesirable properties. The mathematical study of these questions is usually in the more powerful context of weak solutions.

An example of pathological behavior is the sequence (depending upon n) of Cauchy problems for the Laplace equation,

$$\frac{\partial^2 u}{\partial x^2}+\frac{\partial^2 u}{\partial y^2}=0,$$

with boundary conditions,

$$u(x,0) = 0,$$
$$\frac{\partial u}{\partial y}(x,0) = \frac{\sin nx}{n},$$

where n is an integer. The derivative of u with respect to y approaches zero uniformly in x as n increases, but the solution is:

$$u(x,y) = \frac{\sinh ny \sin nx}{n^2}.$$

This solution approaches infinity if nx is not an integer multiple of π for any non-zero value of y. The Cauchy problem for the Laplace equation is called *ill-posed* or *not well-posed*, since the solution does not continuously depend on the data of the problem. Such ill-posed problems are not usually satisfactory for physical applications.

The existence of solutions for the Navier–Stokes equations, a partial differential equation, is part of one of the Millennium Prize Problems.

Notation

In PDEs, it is common to denote partial derivatives using subscripts. That is:

$$u_x = \frac{\partial u}{\partial x}$$

$$u_{xx} = \frac{\partial^2 u}{\partial x^2}$$

$$u_{xy} = \frac{\partial^2 u}{\partial y \partial x} = \frac{\partial}{\partial y}\left(\frac{\partial u}{\partial x}\right).$$

Especially in physics, del or nabla (∇) is often used to denote spatial derivatives, and $\dot{u}$, $\ddot{u}$ for time derivatives. For example, the wave equation (described below) can be written as,

$$\ddot{u} = c^2 \nabla^2 u$$

or

$$\ddot{u} = c\ \Delta u$$

where Δ is the Laplace operator.

Classification

Some linear, second-order partial differential equations can be classified as parabolic, hyperbolic and elliptic. Others, such as the Euler–Tricomi equation, have different types in different regions. The classification provides a guide to appropriate initial and boundary conditions and to the smoothness of the solutions.

FIRST-ORDER PARTIAL DIFFERENTIAL EQUATIONS

In mathematics, a first-order partial differential equation is a partial differential equation that involves only first derivatives of the unknown function of *n* variables. The equation takes the form,

$$F(x_1,\ldots,x_n,u,u_{x_1},\ldots u_{x_n})=0.$$

Such equations arise in the construction of characteristic surfaces for hyperbolic partial differential equations, in the calculus of variations, in some geometrical problems, and in simple models for gas dynamics whose solution involves the method of characteristics. If a family of solutions of a single first-order partial differential equation can be found, then additional solutions may be obtained by forming envelopes of solutions in that family. In a related procedure, general solutions may be obtained by integrating families of ordinary differential equations.

General Solution and Complete Integral

The *general solution* to the first order partial differential equation is a solution which contains an arbitrary function. But, the solution to the first order partial differential equations with as many arbitrary constants as the number of independent variables is called the complete integral. The following n-parameter family of solutions,

$$u=\phi(x_1,x_2,\ldots,x_n,a_1,a_2,\ldots,a_n)$$

is a complete integral if $\det|\phi_{x_i a_j}|\neq 0$.

Characteristic Surfaces for the Wave Equation

Characteristic surfaces for the wave equation are level surfaces for solutions of the equation:

$$u_t^2=c^2\left(u_x^2+u_y^2+u_z^2\right).$$

There is little loss of generality if we set $u_t = 1$: in that case u satisfies,

$$u_x^2 + u_y^2 + u_z^2 = \frac{1}{c^2}.$$

In vector notation, let:

$$\vec{x} = (x,y,z) \quad \text{and} \quad \vec{p} = (u_x, u_y, u_z).$$

A family of solutions with planes as level surfaces is given by,

$$u(\vec{x}) = \vec{p}\cdot(\vec{x} - \overrightarrow{x_0}),$$

where,

$$|\vec{p}| = \frac{1}{c}, \text{ and } \overrightarrow{x_0} \text{ is arbitrary.}$$

If x and x_0 are held fixed, the envelope of these solutions is obtained by finding a point on the sphere of radius $1/c$ where the value of u is stationary. This is true if $\vec{p}$ is parallel to $\vec{x} - \overrightarrow{x_0}$. Hence the envelope has equation,

$$u(\vec{x}) = \pm\frac{1}{c}|\vec{x} - \overrightarrow{x_0}|.$$

These solutions correspond to spheres whose radius grows or shrinks with velocity c. These are light cones in space-time.

The initial value problem for this equation consists in specifying a level surface S where u=0 for t=0. The solution is obtained by taking the envelope of all the spheres with centers on S, whose radii grow with velocity c. This envelope is obtained by requiring that,

$$\frac{1}{c}|\vec{x} - \overrightarrow{x_0}| \text{ is stationary for } \overrightarrow{x_0} \in S.$$

This condition will be satisfied if $|\vec{x} - \overrightarrow{x_0}|$ is normal to S. Thus the envelope corresponds to motion with velocity c along each normal to S. This is the Huygens' construction of wave fronts: each point on S emits a spherical wave at time t=0, and the wave front at a later time t is the envelope of these spherical waves. The normals to S are the light rays.

Two-dimensional Theory

The notation is relatively simple in two space dimensions, but the main ideas generalize to higher dimensions. A general first-order partial differential equation has the form

$$F(x,y,u,p,q) = 0,$$

where,

$$p = u_x, q = u_y.$$

A complete integral of this equation is a solution φ(*x*,*y*,*u*) that depends upon two parameters *a* and *b*. (There are *n* parameters required in the *n*-dimensional case.) An envelope of such solutions is obtained by choosing an arbitrary function *w*, setting *b*=*w*(*a*), and determining *A*(*x*,*y*,*u*) by requiring that the total derivative,

$$\frac{d\varphi}{da} = \varphi_a(x,y,u,A,w(A)) + w'(A)\varphi_b(x,y,u,A,w(A)) = 0.$$

In that case, a solution u_w is also given by,

$$u_w = \phi(x,y,u,A,w(A))$$

Each choice of the function *w* leads to a solution of the PDE. A similar process led to the construction of the light cone as a characteristic surface for the wave equation.

If a complete integral is not available, solutions may still be obtained by solving a system of ordinary equations. To obtain this system, first note that the PDE determines a cone (analogous to the light cone) at each point: if the PDE is linear in the derivatives of *u* (it is quasi-linear), then the cone degenerates into a line. In the general case, the pairs (*p*,*q*) that satisfy the equation determine a family of planes at a given point:

$$u - u_0 = p(x - x_0) + q(y - y_0),$$

where,

$$F(x_0, y_0, u_0, p, q) = 0.$$

The envelope of these planes is a cone, or a line if the PDE is quasi-linear. The condition for an envelope is,

$$F_p\,dp + F_q\,dq = 0,$$

where F is evaluated at (x_0, y_0, u_0, p, q), and *dp* and *dq* are increments of *p* and *q* that satisfy *F*=0. Hence the generator of the cone is a line with direction,

$$dx : dy : du = F_p : F_q : (pF_p + qF_q).$$

This direction corresponds to the light rays for the wave equation. To integrate differential equations along these directions, we require increments for *p* and *q* along the ray. This can be obtained by differentiating the PDE:

$$F_x + F_u p + F_p p_x + F_q p_y = 0,$$

$$F_y + F_u q + F_p q_x + F_q q_y = 0,$$

Therefore the ray direction in (x, y, u, p, q) space is,

$$dx : dy : du : dp : dq = F_p : F_q : (pF_p + qF_q) : (-F_x - F_u p) : (-F_y - F_u q).$$

The integration of these equations leads to a ray conoid at each point (x_0, y_0, u_0). General solutions of the PDE can then be obtained from envelopes of such conoids.

SECOND-ORDER PARTIAL DIFFERENTIAL EQUATIONS

Assuming $u_{xy} = u_{yx}$, the general linear second-order PDE in two independent variables has the form,

$$Au_{xx} + 2Bu_{xy} + Cu_{yy} + \cdots \text{(lower order terms)} = 0,$$

where the coefficients *A*, *B*, *C*... may depend upon *x* and *y*. If $A^2 + B^2 + C^2 > 0$ over a region of the *xy*-plane, the PDE is second-order in that region. This form is analogous to the equation for a conic section:

$$Ax^2 + 2Bxy + Cy^2 + \cdots = 0.$$

More precisely, replacing ∂_x by *X*, and likewise for other variables (formally this is done by a Fourier transform), converts a constant-coefficient PDE into a polynomial of the same degree, with the top degree (a homogeneous polynomial, here a quadratic form) being most significant for the classification.

Just as one classifies conic sections and quadratic forms into parabolic, hyperbolic, and elliptic based on the discriminant $B^2 - 4AC$, the same can be done for a second-order PDE at a given point. However, the discriminant in a PDE is given by $B^2 - AC$ due to the convention of the *xy* term being 2*B* rather than *B*; formally, the discriminant (of the associated quadratic form) is $(2B)^2 - 4AC = 4(B^2 - AC)$, with the factor of 4 dropped for simplicity.

- $B^2 - AC < 0$ (elliptic partial differential equation): Solutions of elliptic PDEs are as smooth as the coefficients allow, within the interior of the region where the equation and solutions are defined. For example, solutions of Laplace's equation are analytic within the domain where they are defined, but solutions may assume boundary values that are not smooth. The motion of a fluid at subsonic speeds can be approximated with elliptic PDEs, and the Euler–Tricomi equation is elliptic where $x < 0$.

- $B^2 - AC = 0$ (parabolic partial differential equation): Equations that are parabolic at every point can be transformed into a form analogous to the heat equation by a change of independent variables. Solutions smooth out as the transformed time variable increases. The Euler–Tricomi equation has parabolic type on the line where $x = 0$.
- $B^2 - AC > 0$ (hyperbolic partial differential equation): hyperbolic equations retain any discontinuities of functions or derivatives in the initial data. An example is the wave equation. The motion of a fluid at supersonic speeds can be approximated with hyperbolic PDEs, and the Euler–Tricomi equation is hyperbolic where $x > 0$.

If there are n independent variables $x_1, x_2, \ldots x_n$, a general linear partial differential equation of second order has the form,

$$Lu = \sum_{i=1}^{n}\sum_{j=1}^{n} a_{i,j}\frac{\partial^2 u}{\partial x_i \partial x_j} \quad \text{plus lower-order terms} = 0.$$

The classification depends upon the signature of the eigenvalues of the coefficient matrix $a_{i,j}$.

- Elliptic: The eigenvalues are all positive or all negative.
- Parabolic: The eigenvalues are all positive or all negative, save one that is zero.
- Hyperbolic: There is only one negative eigenvalue and all the rest are positive, or there is only one positive eigenvalue and all the rest are negative.
- Ultrahyperbolic: There is more than one positive eigenvalue and more than one negative eigenvalue, and there are no zero eigenvalues. There is only a limited theory for ultrahyperbolic equations.

Elliptic Partial Differential Equation

Second order linear partial differential equations (PDEs) are classified as either elliptic, hyperbolic, or parabolic. Any second order linear PDE in two variables can be written in the form

$$Au_{xx} + 2Bu_{xy} + Cu_{yy} + Du_x + Eu_y + Fu + G = 0,$$

where A, B, C, D, E, F, and G are functions of x and y and where $u_x = \frac{\partial u}{\partial x}$ and similarly for $u_{xx}, u_y, u_{yy}, u_{xy}$. A PDE written in this form is elliptic if,

$$B^2 - AC < 0,$$

with this naming convention inspired by the equation for a planar ellipse.

The simplest nontrivial examples of elliptic PDE's are the Laplace equation $\Delta u = u_{xx} + u_{yy} = 0$, and the Poisson equation, $\Delta u = u_{xx} + u_{yy} = f(x,y)$. In a sense, any other elliptic PDE in two variables can be considered to be a generalization of one of these equations, as it can always be put into the canonical form:

$$u_{xx} + u_{yy} + (\text{lower-order terms}) = 0$$

through a change of variables.

Qualitative Behavior

Elliptic equations have no real characteristic curves, curves along which it is not possible to eliminate at least one second derivative of u from the conditions of the Cauchy problem. Since characteristic curves are the only curves along which solutions to partial differential equations with smooth parameters can have discontinuous derivatives, solutions to elliptic equations cannot have discontinuous derivatives anywhere. This means elliptic equations are well suited to describe equilibrium states, where any discontinuities have already been smoothed out. For instance, we can obtain Laplace's equation from the heat equation $u_t = \Delta u$ by setting $u_t = 0$. This means that Laplace's equation describes a steady state of the heat equation.

In parabolic and hyperbolic equations, characteristics describe lines along which information about the initial data travels. Since elliptic equations have no real characteristic curves, there is no meaningful sense of information propagation for elliptic equations. This makes elliptic equations better suited to describe static, rather than dynamic, processes.

Derivation of Canonical Form

We derive the canonical form for elliptic equations in two variables, $u_{xx} + u_{xy} + u_{yy} + (\text{lower-order terms}) = 0$.

$$\xi = \xi(x,y) \text{ and } \eta = \eta(x,y).$$

If $u(\xi,\eta) = u[\xi(x,y),\eta(x,y)]$, applying the chain rule once gives

$$u_x = u_\xi \xi_x + u_\eta \eta_x \text{ and } u_y = u_\xi \xi_y + u_\eta \eta_y,$$

a second application gives,

$$u_{xx} = u_{\xi\xi}\xi^2{}_x + u_{\eta\eta}\eta^2{}_x + 2u_{\xi\eta}\xi_x\eta_x + u_\xi\xi_{xx} + u_\eta\eta_{xx},$$

$$u_{yy} = u_{\xi\xi}\xi^2{}_y + u_{\eta\eta}\eta^2{}_y + 2u_{\xi\eta}\xi_y\eta_y + u_\xi\xi_{yy} + u_\eta\eta_{yy}, \text{ and}$$

$$u_{xy} = u_{\xi\xi}\xi_x\xi_y + u_{\eta\eta}\eta_x\eta_y + u_{\xi\eta}(\xi_x\eta_y + \xi_y\eta_x) + u_\xi\xi_{xy} + u_\eta\eta_{xy}.$$

We can replace our PDE in x and y with an equivalent equation in ξ and η

$$au_{\xi\xi} + 2bu_{\xi\eta} + cu_{\eta\eta} + \text{(lower-order terms)} = 0,$$

where,

$$a = A\xi_x^{\ 2} + 2B\xi_x\xi_y + C\xi_y^{\ 2},$$

$$b = 2A\xi_x\eta_x + 2B(\xi_x\eta_y + \xi_y\eta_x) + 2C\xi_y\eta_y, \text{ and}$$

$$c = A\eta_x^{\ 2} + 2B\eta_x\eta_y + C\eta_y^{\ 2}.$$

To transform our PDE into the desired canonical form, we seek ξ and η such that $a = c$ and $b = 0$. This gives us the system of equations,

$$a - c = A(\xi_x^{\ 2} - \eta_x^{\ 2}) + 2B(\xi_x\xi_y - \eta_x\eta_y) + C(\xi_y^{\ 2} - \eta_y^{\ 2}) = 0$$

$$b = 0 = 2A\xi_x\eta_x + 2B(\xi_x\eta_y + \xi_y\eta_x) + 2C\xi_y\eta_y,$$

Adding i times the second equation to the first and setting $\phi = \xi + i\eta$ gives the quadratic equation,

$$A\phi_x^{\ 2} + 2B\phi_x\phi_y + C\phi_y^{\ 2} = 0.$$

Since the discriminant $B^2 - AC < 0$, this equation has two distinct solutions,

$$\phi_x, \phi_y = \frac{B \pm i\sqrt{AC - B^2}}{A}$$

which are complex conjugates. Choosing either solution, we can solve for $\phi(x,y)$, and recover ξ and η with the transformations $\xi = \operatorname{Re}\phi$ and $\eta = \operatorname{Im}\phi$. Since η and ξ will satisfy $a - c = 0$ and $b = 0$, so with a change of variables from x and y to η and ξ will transform the PDE

$$Au_{xx} + 2Bu_{xy} + Cu_{yy} + Du_x + Eu_y + Fu + G = 0,$$

into the canonical form,

$$u_{\xi\xi} + u_{\eta\eta} + \text{(lower-order terms)} = 0,$$

as desired.

In Higher Dimensions

A general second order partial differential equation in *n* variables takes the form:

$$\sum_{i=1}^{n}\sum_{j=1}^{n} a_{i,j}\frac{\partial^2 u}{\partial x_i \partial x_j} + (\text{lower-order terms}) = 0.$$

This equation is considered elliptic if there are no characteristic surfaces, i.e. surfaces along which it is not possible to eliminate at least one second derivative of *u* from the conditions of the Cauchy problem.

Unlike the two dimensional case, this equation cannot in general be reduced to a simple canonical form.

Parabolic Partial Differential Equation

A parabolic partial differential equation is a type of partial differential equation (PDE). Parabolic PDEs are used to describe a wide variety of time-dependent phenomena, including heat conduction, particle diffusion, and pricing of derivative investment instruments.

To define the simplest kind of parabolic PDE, consider a real-valued function $u(x,y)$ of two independent real variables, x and y. A second-order, linear, constant-coefficient PDE for u takes the form,

$$Au_{xx} + 2Bu_{xy} + Cu_{yy} + Du_x + Eu_y + F = 0,$$

and this PDE is classified as being parabolic if the coefficients satisfy the condition

$$B^2 - AC = 0.$$

Usually x represents one-dimensional position and y represents time, and the PDE is solved subject to prescribed initial and boundary conditions.

The name "parabolic" is used because the assumption on the coefficients is the same as the condition for the analytic geometry equation $Ax^2 + 2Bxy + Cy^2 + Dx + Ey + F = 0$ to define a planar parabola.

The basic example of a parabolic PDE is the one-dimensional heat equation,

$$u_t = \alpha u_{xx},$$

where $u(x,t)$ is the temperature at time t and at position x along a thin rod, and α is a positive constant (the *thermal diffusivity*). The symbol u_t signifies the partial derivative of u with respect to the time variable t, and similarly u_{xx} is the second partial derivative with respect to x. For this example, t plays the role of in the general second-order linear PDE: $A = \alpha$, $E = -1$, and the other coefficients are zero.

The heat equation says, roughly, that temperature at a given time and point rises or falls at a rate proportional to the difference between the temperature at that point and the average temperature near that point. The quantity u_{xx} measures how far off the temperature is from satisfying the mean value property of harmonic functions.

The concept of a parabolic PDE can be generalized in several ways. For instance, the flow of heat through a material body is governed by the three-dimensional heat equation,

$$u_t = \alpha \Delta u,$$

where,

$$\Delta u := \frac{\partial^2 u}{\partial x^2} + \frac{\partial^2 u}{\partial y^2} + \frac{\partial^2 u}{\partial z^2}$$

denotes the Laplace operator acting on u. This equation is the prototype of a multi-dimensional parabolic PDE.

Noting that $-\Delta$ is an elliptic operator suggests a broader definition of a parabolic PDE:

$$u_t = -Lu,$$

where L is a second-order elliptic operator (implying that L must be positive; a case where $u_t = +Lu$ is considered below).

A system of partial differential equations for a vector u can also be parabolic. For example, such a system is hidden in an equation of the form,

$$\nabla \cdot (a(x) \nabla u(x)) + b(x)^T \nabla u(x) + cu(x) = f(x)$$

if the matrix-valued function $a(x)$ has a kernel of dimension 1.

Parabolic PDEs can also be nonlinear. For example, Fisher's equation is a nonlinear PDE that includes the same diffusion term as the heat equation but incorporates a linear growth term and a nonlinear decay term.

Solution:

Under broad assumptions, an initial/boundary-value problem for a linear parabolic PDE has a solution for all time. The solution $u(x,t)$, as a function of x for a fixed time $t > 0$, is generally smoother than the initial data $u(x,0) = u_0(x)$.

For a nonlinear parabolic PDE, a solution of an initial/boundary-value problem might explode in a singularity within a finite amount of time. It can be difficult to determine whether a solution exists for all time, or to understand the singularities that do arise. Such interesting questions arise in the solution of the Poincaré conjecture via Ricci flow.

Backward Parabolic Equation

One occasionally encounters a so-called backward parabolic PDE, which takes the form.

An initial-value problem for the backward heat equation,

$$\begin{cases} u_t = -\Delta u & \text{on } \Omega\times(0,T), \\ u = 0 & \text{on } \partial\Omega\times(0,T), \\ u = f & \text{on } \Omega\times\{0\}. \end{cases}$$

is equivalent to a final-value problem for the ordinary heat equation,

$$\begin{cases} u_t = \Delta u & \text{on } \Omega\times(0,T), \\ u = 0 & \text{on } \partial\Omega\times(0,T), \\ u = f & \text{on } \Omega\times\{T\}. \end{cases}$$

An initial/boundary-value problem for a backward parabolic PDE is usually not well-posed (solutions often grow unbounded in finite time, or even fail to exist). Nonetheless, these problems are important for the study of the reflection of singularities of solutions to various other PDEs. Moreover, they arise in the pricing problem for certain financial instruments.

Hyperbolic Partial Differential Equation

In mathematics, a hyperbolic partial differential equation of order n is a partial differential equation (PDE) that, roughly speaking, has a well-posed initial value problem for the first $n - 1$ derivatives. More precisely, the Cauchy problem can be locally solved for arbitrary initial data along any non-characteristic hypersurface. Many of the equations of mechanics are hyperbolic, and so the study of hyperbolic equations is of substantial contemporary interest. The model hyperbolic equation is the wave equation. In one spatial dimension, this is,

$$\frac{\partial^2 u}{\partial t^2} = c^2 \frac{\partial^2 u}{\partial x^2}$$

The equation has the property that, if u and its first time derivative are arbitrarily specified initial data on the line $t = 0$ (with sufficient smoothness properties), then there exists a solution for all time t.

The solutions of hyperbolic equations are "wave-like". If a disturbance is made in the initial data of a hyperbolic differential equation, then not every point of space feels the disturbance at once. Relative to a fixed time coordinate, disturbances have a finite propagation speed. They travel along the characteristics of the equation. This feature qualitatively distinguishes hyperbolic equations from elliptic partial differential equations and parabolic partial differential equations. A perturbation of the initial (or boundary) data of an elliptic or parabolic equation is felt at once by essentially all points in the domain.

Although the definition of hyperbolicity is fundamentally a qualitative one, there are precise criteria that depend on the particular kind of differential equation under consideration. There is a well-developed theory for linear differential operators, due to Lars Gårding, in the context of microlocal analysis. Nonlinear differential equations are hyperbolic if their linearizations are hyperbolic in the sense of Gårding. There is a somewhat different theory for first order systems of equations coming from systems of conservation laws.

A partial differential equation is hyperbolic at a point *P* provided that the Cauchy problem is uniquely solvable in a neighborhood of *P* for any initial data given on a non-characteristic hypersurface passing through *P*. Here the prescribed initial data consist of all (transverse) derivatives of the function on the surface up to one less than the order of the differential equation.

Examples:

By a linear change of variables, any equation of the form,

$$A\frac{\partial^2 u}{\partial x^2}+2B\frac{\partial^2 u}{\partial x\partial y}+C\frac{\partial^2 u}{\partial y^2}+(\text{lower order derivative terms})=0$$

with,

$$B^2-AC>0$$

can be transformed to the wave equation, apart from lower order terms which are inessential for the qualitative understanding of the equation. This definition is analogous to the definition of a planar hyperbola.

The one-dimensional wave equation:

$$\frac{\partial^2 u}{\partial t^2}-c^2\frac{\partial^2 u}{\partial x^2}=0$$

is an example of a hyperbolic equation. The two-dimensional and three-dimensional wave equations also fall into the category of hyperbolic PDE. This type of second-order hyperbolic partial differential equation may be transformed to a hyperbolic system of first-order differential equations.

Hyperbolic System of Partial Differential Equations

The following is a system of s first order partial differential equations for s unknown functions $\vec{u}=(u_1,\ldots,u_s)$, $\vec{u}=\vec{u}(\vec{x},t)$, where $\vec{x}\in\mathbb{R}^d$:

$$(*)\quad \frac{\partial\vec{u}}{\partial t}+\sum_{j=1}^{d}\frac{\partial}{\partial x_j}\vec{f^j}(\vec{u})=0,$$

where $\vec{f^j}\in C^1(\mathbb{R}^s,\mathbb{R}^s), j=1,\ldots,d$ are once continuously differentiable functions, nonlinear in general.

Next, for each $\vec{f^j}$ define the $s\times s$ Jacobian matrix:

$$A^j:=\begin{pmatrix}\frac{\partial f_1^j}{\partial u_1} & \cdots & \frac{\partial f_1^j}{\partial u_s}\\ \vdots & \ddots & \vdots\\ \frac{\partial f_s^j}{\partial u_1} & \cdots & \frac{\partial f_s^j}{\partial u_s}\end{pmatrix},\ \text{for } j=1,\ldots,d.$$

The system $(*)$ is hyperbolic if for all $\alpha_1,\ldots,\alpha_d\in\mathbb{R}$ the matrix $A:=\alpha_1A^1+\cdots+\alpha_dA^d$ has only real eigenvalues and is diagonalizable.

If the matrix A has *s distinct* real eigenvalues, it follows that it is diagonalizable. In this case the system $(*)$ is called strictly hyperbolic.

If the matrix A is symmetric, it follows that it is diagonalizable and the eigenvalues are real. In this case the system $(*)$ is called symmetric hyperbolic.

Hyperbolic System and Conservation Laws

There is a connection between a hyperbolic system and a conservation law. Consider a hyperbolic system of one partial differential equation for one unknown function $u=u(\vec{x},t)$. Then the system $(*)$ has the form,

$$(**)\quad \frac{\partial u}{\partial t}+\sum_{j=1}^{d}\frac{\partial}{\partial x_j}f^j(u)=0.$$

Here, u can be interpreted as a quantity that moves around according to the flux given by $\vec{f}=\left(f^1,\ldots,f^d\right)$. To see that the quantity u is conserved, integrate $(**)$ over a domain Ω

$$\int_\Omega\frac{\partial u}{\partial t}d\Omega+\int_\Omega\nabla\cdot\vec{f}(u)d\Omega=0.$$

If u and $\vec{f}$ are sufficiently smooth functions, we can use the divergence theorem and change the order of the integration and $\partial / \partial t$ to get a conservation law for the quantity u in the general form,

$$\frac{d}{dt}\int_{\Omega} u \, d\Omega + \int_{\partial\Omega} \vec{f}(u) \cdot \vec{n} \, d\Gamma = 0,$$

which means that the time rate of change of u in the domain Ω is equal to the net flux of u through its boundary $\partial\Omega$. Since this is an equality, it can be concluded that u is conserved within Ω.

SEPARABLE PARTIAL DIFFERENTIAL EQUATION

A separable partial differential equation (PDE) is one that can be broken into a set of separate equations of lower dimensionality (fewer independent variables) by a method of separation of variables. This generally relies upon the problem having some special form or symmetry. In this way, the PDE can be solved by solving a set of simpler PDEs, or even ordinary differential equations (ODEs) if the problem can be broken down into one-dimensional equations.

The most common form of separation of variables is simple separation of variables in which a solution is obtained by assuming a solution of the form given by a product of functions of each individual coordinate. There is a special form of separation of variables called -separation of variables which is accomplished by writing the solution as a particular fixed function of the coordinates multiplied by a product of functions of each individual coordinate. Laplace's equation on $\mathbb{R}^n$ is an example of a partial differential equation which admits solutions through R -separation of variables; in the three-dimensional case this uses 6-sphere coordinates.

Example:

For example, consider the time-independent Schrödinger equation,

$$[-\nabla^2 + V(\mathbf{x})]\psi(\mathbf{x}) = E\psi(\mathbf{x})$$

for the function $\psi(\mathbf{x})$ (in dimensionless units, for simplicity). (Equivalently, consider the inhomogeneous Helmholtz equation.) If the function $V(\mathbf{x})$ in three dimensions is of the form,

$$V(x_1, x_2, x_3) = V_1(x_1) + V_2(x_2) + V_3(x_3),$$

then it turns out that the problem can be separated into three one-dimensional ODEs for functions $\psi_1(x_1)$, $\psi_2(x_2)$, and $\psi_3(x_3)$, and the final solution can be written as $\psi(x) = \psi_1(x_1)\cdot\psi_2(x_2)\cdot\psi_3(x_3)$. (More generally, the separable cases of the Schrödinger equation were enumerated by Eisenhart).

METHOD OF CHARACTERISTICS

In mathematics, the method of characteristics is a technique for solving partial differential equations. Typically, it applies to first-order equations, although more generally the method of characteristics is valid for any hyperbolic partial differential equation. The method is to reduce a partial differential equation to a family of ordinary differential equations along which the solution can be integrated from some initial data given on a suitable hypersurface.

Characteristics of First-order Partial Differential Equation

For a first-order PDE (partial differential equation), the method of characteristics discovers curves (called characteristic curves or just characteristics) along which the PDE becomes an ordinary differential equation (ODE). Once the ODE is found, it can be solved along the characteristic curves and transformed into a solution for the original PDE.

For the sake of simplicity, we confine our attention to the case of a function of two independent variables x and y for the moment. Consider a quasilinear PDE of the form,

$$a(x,y,z)\frac{\partial z}{\partial x} + b(x,y,z)\frac{\partial z}{\partial y} = c(x,y,z).$$

Suppose that a solution z is known, and consider the surface graph $z = z(x,y)$ in $\mathbf{R}^3$. A normal vector to this surface is given by,

$$\left(\frac{\partial z}{\partial x}(x,y), \frac{\partial z}{\partial y}(x,y), -1\right).$$

As a result, equation $a(x,y,z)\ldots = c(x,y,z)$ is equivalent to the geometrical statement that the vector field,

$$(a(x,y,z), b(x,y,z), c(x,y,z))$$

is tangent to the surface $z = z(x,y)$ at every point, for the dot product of this vector field with the above normal vector is zero. In other words, the graph of the solution must be a union of integral curves of this vector field. These integral curves are called the characteristic curves of the original partial differential equation and are given by the Lagrange-Charpit equations:

$$\frac{dx}{dt} = a(x,y,z)$$
$$\frac{dy}{dt} = b(x,y,z)$$
$$\frac{dz}{dt} = c(x,y,z).$$

A parametrization invariant form of the Lagrange-Charpit equations is:

$$\frac{dx}{a(x,y,z)} = \frac{dy}{b(x,y,z)} = \frac{dz}{c(x,y,z)}.$$

Linear and Quasilinear Cases

Consider now a PDE of the form,

$$\sum_{i=1}^{n} a_i(x_1,\dots,x_n,u)\frac{\partial u}{\partial x_i} = c(x_1,\dots,x_n,u).$$

For this PDE to be linear, the coefficients a_i may be functions of the spatial variables only, and independent of u. For it to be quasilinear, a_i may also depend on the value of the function, but not on any derivatives.

For a linear or quasilinear PDE, the characteristic curves are given parametrically by,

$$(x_1,\dots,x_n,u) = (x_1(s),\dots,x_n(s),u(s))$$

$$u(\mathbf{X}(s)) = U(s)$$

such that the following system of ODEs is satisfied,

$$\frac{dx_i}{ds} = a_i(x_1,\dots,x_n,u)$$

$$\frac{du}{ds} = c(x_1,\dots,x_n,u).$$

Equations above give the characteristics of the PDE.

Quasi-linear Case

In the Quasi-Linear case, the use of the method of characteristics is justified by Grönwall's Inequality. The above equation may be written as,

$$a(\mathbf{x},u)\cdot\nabla u(\mathbf{x}) = c(\mathbf{x},u)$$

We must distinguish between the solutions to the ODE and the solutions to the PDE, which we do not know are equal *a priori*. Letting capital letters be the solutions to the ODE we find,

$$\mathbf{X}'(s) = a(\mathbf{X}(s),U(s))$$

$$U'(s) = c(\mathbf{X}(s),U(s))$$

Examining $\Delta(s) = |u(\mathbf{X}(s)) - U(s)|^2$, we find, upon differentiating that,

$$\Delta'(s) = 2\big(u(\mathbf{X}(s)) - U(s)\big)\big(\mathbf{X}'(s)\cdot\nabla u(\mathbf{X}(s)) - U'(s)\big)$$

which is the same as,

$$\Delta'(s) = 2\big(u(\mathbf{X}(s)) - U(s)\big)\big(a(\mathbf{X}(s),U(s))\cdot\nabla u(\mathbf{X}(s)) - c(\mathbf{X}(s),U(s))\big)$$

We cannot conclude the above is 0 as we would like, since the PDE only guarantees us that this relationship is satisfied for $u(\mathbf{x})$, $a(\mathbf{x},u)\cdot\nabla u(\mathbf{x}) = c(\mathbf{x},u)$, and we do not yet know that $U(s) = u(\mathbf{X}(s))$.

However, we can see that,

$$\Delta'(s) = 2\big(u(\mathbf{X}(s)) - U(s)\big)\Big(a(\mathbf{X}(s),U(s))\cdot\nabla u(\mathbf{X}(s)) - c(\mathbf{X}(s),U(s)) -$$

$$\big(a(\mathbf{X}(s),u(\mathbf{X}(s)))\cdot\nabla u(\mathbf{X}(s)) - c(\mathbf{X}(s),u(\mathbf{X}(s)))\big)\Big)$$

since by the PDE, the last term is 0. This equals,

$$\Delta'(s) = 2\big(u(\mathbf{X}(s)) - U(s)\big)\Big(\big(a(\mathbf{X}(s),U(s)) - a(\mathbf{X}(s),u(\mathbf{X}(s)))\big)\cdot$$

$$\nabla u(\mathbf{X}(s)) - \big(c(\mathbf{X}(s),U(s)) - c(\mathbf{X}(s),u(\mathbf{X}(s)))\big)\Big)$$

By the triangle inequality, we have,

$$|\Delta'(s)| = 2\big|u(\mathbf{X}(s)) - U(s)\big|\Big(\big\|a(\mathbf{X}(s),U(s)) - a(\mathbf{X}(s),u(\mathbf{X}(s)))\big\|$$

$$\big\|\nabla u(\mathbf{X}(s))\big\| + \big|c(\mathbf{X}(s),U(s)) - c(\mathbf{X}(s),u(\mathbf{X}(s)))\big|\Big)$$

Assuming $\mathbf{a}, c$ are at least C^1, we can bound this for small times. Choose a neighborhood Ω around $\mathbf{X}(0), U(0)$ small enough such that $\mathbf{a}, c$ are Locally Lipschitz. By continuity, $(\mathbf{X}(s), U(s))$ will remain in Ω for small enough s. Since $U(0) = u(\mathbf{X}(0))$, we also have that $(\mathbf{X}(s), u(\mathbf{X}(s)))$ will be in Ω for small enough s by continuity. So, $(\mathbf{X}(s), U(s)) \in \Omega$ and $(\mathbf{X}(s), u(\mathbf{X}(s))) \in \Omega$ for $s \in [0, s_0]$. Additionally, $\|\nabla u(\mathbf{X}(s))\| \le M$ for some $M \in \mathbb{R}$ for $s \in [0, s_0]$ by compactness. From this, we find the above is bounded as,

$$|\Delta'(s)| \le C|u(\mathbf{X}(s)) - U(s)|^2 = C|\Delta(s)|$$

for some $C \in \mathbb{R}$. It is a straightforward application of Grönwall's Inequality to show that since $\Delta(0) = 0$ we have $\Delta(s) = 0$ for as long as this inequality holds. We have some interval $[0, \epsilon)$ such that $u(\mathbf{X}(s)) = U(s)$ in this interval. Choose the largest ϵ such that this is true. Then, by continuity, $U(\epsilon) = u(\mathbf{X}(\epsilon))$. Provided the ODE still has a solution in some interval after ϵ, we can repeat the argument above to find that $u(\mathbf{X}(s)) = U(s)$ in a larger interval. Thus, so long as the ODE has a solution, we have $u(\mathbf{X}(s)) = U(s)$.

Fully Nonlinear Case

Consider the partial differential equation,

$$F(x_1, \ldots, x_n, u, p_1, \ldots, p_n) = 0$$

where the variables p_i are shorthand for the partial derivatives,

$$p_i = \frac{\partial u}{\partial x_i}.$$

Let $(x_i(s), u(s), p_i(s))$ be a curve in $\mathbf{R}^{2n+1}$. Suppose that u is any solution, and that,

$$u(s) = u(x_1(s), \ldots, x_n(s)).$$

Along a solution, differentiating equation $F(x_1, \ldots, x_n, u, p_1, \ldots, p_n) = 0$ with respect to s gives,

$$\sum_i \left(F_{x_i} + F_u p_i\right)\dot{x}_i + \sum_i F_{p_i}\dot{p}_i = 0$$

$$\dot{u} - \sum_i p_i \dot{x}_i = 0$$

$$\sum_i (\dot{x}_i dp_i - \dot{p}_i dx_i) = 0.$$

The second equation follows from applying the chain rule to a solution *u*, and the third follows by taking an exterior derivative of the relation $du - \sum_i p_i \, dx_i = 0$. Manipulating these equations gives,

$$\dot{x}_i = \lambda F_{p_i}, \quad \dot{p}_i = -\lambda(F_{x_i} + F_u p_i), \quad \dot{u} = \lambda \sum_i p_i F_{p_i}$$

where λ is a constant. Writing these equations more symmetrically, one obtains the Lagrange-Charpit equations for the characteristic,

$$\frac{\dot{x}_i}{F_{p_i}} = -\frac{\dot{p}_i}{F_{x_i} + F_u p_i} = \frac{\dot{u}}{\sum p_i F_{p_i}}.$$

Geometrically, the method of characteristics in the fully nonlinear case can be interpreted as requiring that the Monge cone of the differential equation should everywhere be tangent to the graph of the solution.

Example:

As an example, consider the advection equation (this example assumes familiarity with PDE notation, and solutions to basic ODEs).

$$a\frac{\partial u}{\partial x} + \frac{\partial u}{\partial t} = 0$$

where a is constant and u is a function of x and t. We want to transform this linear first-order PDE into an ODE along the appropriate curve; i.e. something of the form,

$$\frac{d}{ds}u(x(s),t(s)) = F(u,x(s),t(s)),$$

where $(x(s),t(s))$ is a characteristic line. First, we find,

$$\frac{d}{ds}u(x(s),t(s)) = \frac{\partial u}{\partial x}\frac{dx}{ds} + \frac{\partial u}{\partial t}\frac{dt}{ds}$$

by the chain rule. Now, if we set $\frac{dx}{ds} = a$ and $\frac{dt}{ds} = 1$ we get,

$$a\frac{\partial u}{\partial x} + \frac{\partial u}{\partial t}$$

which is the left hand side of the PDE we started with. Thus,

$$\frac{d}{ds}u = a\frac{\partial u}{\partial x} + \frac{\partial u}{\partial t} = 0.$$

So, along the characteristic line $(x(s),t(s))$, the original PDE becomes the ODE $u_s = F(u,x(s),t(s)) = 0$. That is to say that along the characteristics, the solution is constant. Thus, $u(x_s,t_s) = u(x_0,0)$ where (x_s,t_s) and $(x_0,0)$ lie on the same characteristic. Therefore, to determine the general solution, it is enough to find the characteristics by solving the characteristic system of ODEs:

- $\frac{dt}{ds} = 1$, letting $t(0) = 0$ we know $t = s$,
- $\frac{dx}{ds} = a$, letting $x(0) = x_0$ we know $x = as + x_0 = at + x_0$,
- $\frac{du}{ds} = 0$, letting $u(0) = f(x_0)$ we know $u(x(t),t) = f(x_0) = f(x - at)$.

In this case, the characteristic lines are straight lines with slope a, and the value of u remains constant along any characteristic line.

Characteristics of Linear Differential Operators

Let *X* be a differentiable manifold and *P* a linear differential operator,

$$P : C^\infty(X) \to C^\infty(X)$$

of order *k*. In a local coordinate system x^i,

$$P = \sum_{|\alpha| \le k} P^\alpha(x) \frac{\partial}{\partial x^\alpha}$$

in which α denotes a multi-index. The principal symbol of *P*, denoted σ_P, is the function on the cotangent bundle T^*X defined in these local coordinates by,

$$\sigma_P(x,\xi) = \sum_{|\alpha| = k} P^\alpha(x)\xi_\alpha$$

where the ξ_i are the fiber coordinates on the cotangent bundle induced by the coordinate differentials dx^i. Although this is defined using a particular coordinate system, the transformation law relating the ξ_i and the x^i ensures that σ_P is a well-defined function on the cotangent bundle.

The function σ_P is homogeneous of degree *k* in the ξ variable. The zeros of σ_P, away from the zero section of T^*X, are the characteristics of *P*. A hypersurface of *X* defined by the equation $F(x) = c$ is called a characteristic hypersurface at *x* if,

$$\sigma_P\left(x, dF(x)\right) = 0.$$

Invariantly, a characteristic hypersurface is a hypersurface whose conormal bundle is in the characteristic set of *P*.

Qualitative Analysis of Characteristics

Characteristics are also a powerful tool for gaining qualitative insight into a PDE.

One can use the crossings of the characteristics to find shock waves for potential flow in a compressible fluid. Intuitively, we can think of each characteristic line implying a solution to u along itself. Thus, when two characteristics cross, the function becomes multi-valued resulting in a non-physical solution. Physically, this contradiction is removed by the formation of a shock wave, a tangential discontinuity or a weak discontinuity and can result in non-potential flow, violating the initial assumptions.

Characteristics may fail to cover part of the domain of the PDE. This is called a rarefaction, and indicates the solution typically exists only in a weak, i.e. integral equation, sense.

The direction of the characteristic lines indicate the flow of values through the solution, as the example above demonstrates. This kind of knowledge is useful when solving PDEs numerically as it can indicate which finite difference scheme is best for the problem.

FUNDAMENTAL SOLUTION

In mathematics, a fundamental solution for a linear partial differential operator *L* is a formulation in the language of distribution theory of the older idea of a Green's function (although unlike Green's functions, fundamental solutions do not address boundary conditions).

In terms of the Dirac delta "function" $\delta(x)$, a fundamental solution *F* is the solution of the inhomogeneous equation,

$$LF=\delta(x).$$

Here *F* is a priori only assumed to be a distribution.

This concept has long been utilized for the Laplacian in two and three dimensions. It was investigated for all dimensions for the Laplacian by Marcel Riesz.

The existence of a fundamental solution for any operator with constant coefficients — the most important case, directly linked to the possibility of using convolution to solve an arbitrary right hand side — was shown by Bernard Malgrange and Leon Ehrenpreis.

Example:

Consider the following differential equation $Lf = \sin(x)$ with,

$$L=\frac{d^2}{dx^2}.$$

The fundamental solutions can be obtained by solving $LF = \delta(x)$, explicitly,

$$\frac{d^2}{dx^2}F(x)=\delta(x).$$

Since for the Heaviside function H we have,

$$\frac{d}{dx}H(x)=\delta(x),$$

there is a solution

$$\frac{d}{dx}F(x)=H(x)+C.$$

Here C is an arbitrary constant introduced by the integration. For convenience, set $C = -1/2$.

After integrating $\frac{dF}{dx}$ and choosing the new integration constant as zero,

$$F(x)=xH(x)-\frac{1}{2}x=\frac{1}{2}|x|.$$

Motivation

Once the fundamental solution is found, it is straightforward to find a solution of the original equation, through convolution of the fundamental solution and the desired right hand side.

Fundamental solutions also play an important role in the numerical solution of partial differential equations by the boundary element method.

Consider the operator L and the differential equation mentioned in the example,

$$\frac{d^2}{dx^2}f(x)=\sin(x).$$

We can find the solution $f(x)$ of the original equation by convolution (denoted by an asterisk) of the right-hand side $\sin(x)$ with the fundamental solution $F(x)=\frac{|x|}{2}$:

$$f(x)=(F*\sin)(x):=\int_{-\infty}^{\infty}\frac{1}{2}|x-y|\sin(y)\,dy.$$

This shows that some care must be taken when working with functions which do not have enough regularity (e.g. compact support, L^1 integrability) since, we know that the desired solution is $f(x) = -\sin x$, while the above integral diverges for all x. The two expressions for f are, however, equal as distributions.

An example that more clearly works:

$$\frac{d^2}{dx^2}f(x)=I(x),$$

where I is the characteristic (indicator) function of the unit interval $[0,1]$. In that case, it can be readily verified that the convolution $I*F$ with $F(x)=|x|/2$ is a solution, i.e., has second derivative equal to I.

Proof that the convolution is a solution:

Denote the convolution of functions F and g as $F*g$. Say we are trying to find the solution of $Lf = g(x)$. We want to prove that $F*g$ is a solution of the previous equation, i.e. we want to prove that $L(F*g) = g$. When applying the differential operator, L, to the convolution, it is known that,

$$L(F*g)=(LF)*g,$$

provided L has constant coefficients.

If F is the fundamental solution, the right side of the equation reduces to,

$$\delta*g.$$

But since the delta function is an identity element for convolution, this is simply $g(x)$. Summing up,

$$L(F*g)=(LF)*g=\delta(x)*g(x)=\int_{-\infty}^{\infty}\delta(x-y)g(y)dy=g(x).$$

Therefore, if F is the fundamental solution, the convolution $F*g$ is one solution of $Lf = g(x)$. This does not mean that it is the only solution. Several solutions for different initial conditions can be found.

Fundamental Solutions for Some Partial Differential Equations

The following can be obtained by means of Fourier transform:

Laplace Equation

For the Laplace equation,

$$[-\Delta]\Phi(\mathbf{x},\mathbf{x}')=\delta(\mathbf{x}-\mathbf{x}')$$

the fundamental solutions in two and three dimensions, respectively, are,

$$\Phi_{2D}(\mathbf{x},\mathbf{x}') = -\frac{1}{2\pi}\ln|\mathbf{x}-\mathbf{x}'|, \qquad \Phi_{3D}(\mathbf{x},\mathbf{x}') = \frac{1}{4\pi|\mathbf{x}-\mathbf{x}'|}.$$

Screened Poisson Equation

For the screened Poisson equation,

$$[-\Delta + k^2]\Phi(\mathbf{x},\mathbf{x}') = \delta(\mathbf{x}-\mathbf{x}'), \quad k \in \mathbb{R},$$

the fundamental solutions are,

$$\Phi_{2D}(\mathbf{x},\mathbf{x}') = \frac{1}{2\pi}K_0(k|\mathbf{x}-\mathbf{x}'|), \qquad \Phi_{3D}(\mathbf{x},\mathbf{x}') = \frac{\exp(-k|\mathbf{x}-\mathbf{x}'|)}{4\pi|\mathbf{x}-\mathbf{x}'|},$$

where K_0 is a modified Bessel function of the second kind.

In higher dimensions the fundamental solution of the screened Poisson equation is given by the Bessel potential.

Biharmonic Equation

For the Biharmonic equation,

$$[-\Delta^2]\Phi(\mathbf{x},\mathbf{x}') = \delta(\mathbf{x}-\mathbf{x}')$$

the biharmonic equation has the fundamental solutions,

$$\Phi_{2D}(\mathbf{x},\mathbf{x}') = -\frac{|\mathbf{x}-\mathbf{x}'|^2}{8\pi}\ln|x-x'|, \qquad \Phi_{3D}(\mathbf{x},\mathbf{x}') = \frac{|\mathbf{x}-\mathbf{x}'|}{8\pi}.$$

Signal Processing

In signal processing, the analog of the fundamental solution of a differential equation is called the impulse response of a filter.

BOUNDARY VALUE PROBLEM

In mathematics, in the field of differential equations, a boundary value problem is a differential equation together with a set of additional constraints, called the boundary conditions. A solution to a boundary value problem is a solution to the differential equation which also satisfies the boundary conditions.

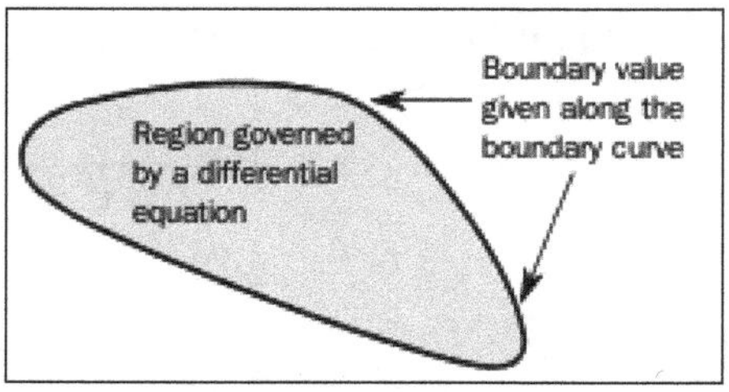

Shows a region where a differential equation is valid and the associated boundary values.

Boundary value problems arise in several branches of physics as any physical differential equation will have them. Problems involving the wave equation, such as the determination of normal modes, are often stated as boundary value problems. A large class of important boundary value problems are the Sturm–Liouville problems. The analysis of these problems involves the eigenfunctions of a differential operator.

To be useful in applications, a boundary value problem should be well posed. This means that given the input to the problem there exists a unique solution, which depends continuously on the input. Much theoretical work in the field of partial differential equations is devoted to proving that boundary value problems arising from scientific and engineering applications are in fact well-posed.

Among the earliest boundary value problems to be studied is the Dirichlet problem, of finding the harmonic functions (solutions to Laplace's equation); the solution was given by the Dirichlet's principle.

Boundary value problems are similar to initial value problems. A boundary value problem has conditions specified at the extremes ("boundaries") of the independent variable in the equation whereas an initial value problem has all of the conditions specified at the same value of the independent variable (and that value is at the lower boundary of the domain, thus the term "initial" value). A boundary value is a data value that corresponds to a minimum or maximum input, internal, or output value specified for a system or component.

For example, if the independent variable is time over the domain [0,1], a boundary value problem would specify values for $y(t)$ at both $t=0$ and $t=1$, whereas an initial value problem would specify a value of $y(t)$ and $y'(t)$ at time $t=0$.

Finding the temperature at all points of an iron bar with one end kept at absolute zero and the other end at the freezing point of water would be a boundary value problem.

If the problem is dependent on both space and time, one could specify the value of the problem at a given point for all time or at a given time for all space.

Concretely, an example of a boundary value (in one spatial dimension) is the problem,

$$y''(x)+y(x)=0$$

to be solved for the unknown function $y(x)$ with the boundary conditions,

$$y(0)=0, y(\pi/2)=2.$$

Without the boundary conditions, the general solution to this equation is,

$$y(x)=A\sin(x)+B\cos(x).$$

From the boundary condition $y(0)=0$ one obtains,

$$0=A\cdot 0+B\cdot 1$$

which implies that $B=0$. From the boundary condition $y(\pi/2)=2$ one finds,

$$2=A\cdot 1$$

and so $A=2$. One sees that imposing boundary conditions allowed one to determine a unique solution, which in this case is,

$$y(x)=2\sin(x).$$

Types of Boundary Value Problems

Boundary Value Conditions

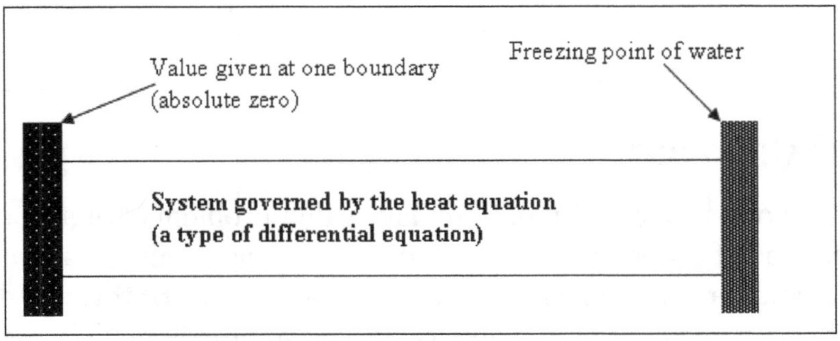

Finding a function to describe the temperature of this idealised 2D rod is a boundary value problem with Dirichlet boundary conditions. Any solution function will both solve the heat equation, and fulfill the boundary conditions of a temperature of 0 K on the left boundary and a temperature of 273.15 K on the right boundary.

A boundary condition which specifies the value of the function itself is a Dirichlet boundary condition, or first-type boundary condition. For example, if one end of an iron rod is held at absolute zero, then the value of the problem would be known at that point in space.

A boundary condition which specifies the value of the normal derivative of the function is a Neumann boundary condition, or second-type boundary condition. For example, if there is a heater at one end of an iron rod, then energy would be added at a constant rate but the actual temperature would not be known.

If the boundary has the form of a curve or surface that gives a value to the normal derivative and the variable itself then it is a Cauchy boundary condition.

Examples:

Summary of boundary conditions for the unknown function, y, constants c_0 and c_1 specified by the boundary conditions, and known scalar functions f and g specified by the boundary conditions.

Name	Form on 1st part of boundary	Form on 2nd part of boundary
Dirichlet	$y = f$	
Neumann	$\frac{\partial y}{\partial n} = f$	
Robin	$c_0 y + c_1 \frac{\partial y}{\partial n} = f$	
Mixed	$y = f$	$c_0 y + c_1 \frac{\partial y}{\partial n} = f$
Cauchy	both $y = f$, and $c_0 \frac{\partial y}{\partial n} = g$	

Differential Operators

Aside from the boundary condition, boundary value problems are also classified according to the type of differential operator involved. For an elliptic operator, one discusses elliptic boundary value problems. For a hyperbolic operator, one discusses hyperbolic boundary value problems. These categories are further subdivided into linear and various nonlinear types.

Applications

Electromagnetic Potential

In electrostatics, a common problem is to find a function which describes the electric potential of a given region. If the region does not contain charge, the potential must be a solution to Laplace's equation (a so-called harmonic function). The boundary conditions in this case are the Interface conditions for electromagnetic fields. If there is no current density in the region, it is also possible to define a magnetic scalar potential using a similar procedure.

NONLINEAR PARTIAL DIFFERENTIAL EQUATION

In mathematics and physics, a nonlinear partial differential equation is a partial differential equation with nonlinear terms. They describe many different physical systems, ranging from gravitation to fluid dynamics, and have been used in mathematics to solve problems such as the Poincaré conjecture and the Calabi conjecture. They are difficult to study: there are almost no general techniques that work for all such equations, and usually each individual equation has to be studied as a separate problem.

Methods for Studying Nonlinear Partial Differential Equations

Existence and Uniqueness of Solutions

A fundamental question for any PDE is the existence and uniqueness of a solution for given boundary conditions. For nonlinear equations these questions are in general very hard: for example, the hardest part of Yau's solution of the Calabi conjecture was the proof of existence for a Monge–Ampere equation.

Singularities

The basic questions about singularities (their formation, propagation, and removal, and regularity of solutions) are the same as for linear PDE, but as usual much harder to study. In the linear case one can just use spaces of distributions, but nonlinear PDEs are not usually defined on arbitrary distributions, so one replaces spaces of distributions by refinements such as Sobolev spaces.

An example of singularity formation is given by the Ricci flow: Richard S. Hamilton showed that while short time solutions exist, singularities will usually form after a finite time. Grigori Perelman's solution of the Poincaré conjecture depended on a deep study of these singularities, where he showed how to continue the solution past the singularities.

Linear Approximation

The solutions in a neighborhood of a known solution can sometimes be studied by linearizing the PDE around the solution. This corresponds to studying the tangent space of a point of the moduli space of all solutions.

Moduli Space of Solutions

Ideally one would like to describe the (moduli) space of all solutions explicitly, and for some very special PDEs this is possible. (In general this is a hopeless problem: it is unlikely that there is any useful description of all solutions of the Navier–Stokes equation for example, as this would involve describing all possible fluid motions.) If the equation has a very large symmetry group, then one is usually only interested in the moduli space of solutions modulo the symmetry group, and this is sometimes a finite-dimensional compact manifold, possibly with singularities; for example, this happens in the case of the Seiberg–Witten equations. A slightly more complicated case is the self dual Yang–Mills equations, when the moduli space is finite-dimensional but not necessarily compact, though it can often be compactified explicitly. Another case when one can sometimes hope to describe all solutions is the case of completely integrable models, when solutions are sometimes a sort of superposition of solitons; this happens e.g. for the Korteweg–de Vries equation.

Exact Solutions

It is often possible to write down some special solutions explicitly in terms of elementary functions (though it is rarely possible to describe all solutions like this). One way of finding such explicit solutions is to reduce the equations to equations of lower dimension, preferably ordinary differential equations, which can often be solved exactly. This can sometimes be done using separation of variables, or by looking for highly symmetric solutions.

Some equations have several different exact solutions.

Lax Pair

If a system of PDEs can be put into Lax pair form,

$$\frac{dL}{dt} = LA - AL$$

then it usually has an infinite number of first integrals, which help to study it.

Euler–Lagrange Equations

Systems of PDEs often arise as the Euler–Lagrange equations for a variational problem. Systems of this form can sometimes be solved by finding an extremum of the original variational problem.

Integrable Systems

PDEs that arise from integrable systems are often the easiest to study, and can sometimes be completely solved. A well-known example is the Korteweg–de Vries equation.

Symmetry

Some systems of PDEs have large symmetry groups. For example, the Yang–Mills equations are invariant under an infinite-dimensional gauge group, and many systems of equations (such as the Einstein field equations) are invariant under diffeomorphisms of the underlying manifold. Any such symmetry groups can usually be used to help study the equations; in particular if one solution is known one can trivially generate more by acting with the symmetry group.

Sometimes equations are parabolic or hyperbolic "modulo the action of some group": for example, the Ricci flow equation is not quite parabolic, but is "parabolic modulo the action of the diffeomorphism group", which implies that it has most of the good properties of parabolic equations.

NUMERICAL METHODS FOR PARTIAL DIFFERENTIAL EQUATIONS

The three most widely used numerical methods to solve PDEs are the finite element method (FEM), finite volume methods (FVM) and finite difference methods (FDM), as well other kind of methods called Meshfree methods, which were made to solve problems where the before mentioned methods are limited. The FEM has a prominent position among these methods and especially its exceptionally efficient higher-order version hp-FEM. Other hybrid versions of FEM and Meshfree methods include the generalized finite element method (GFEM), extended finite element method (XFEM), spectral finite element method (SFEM), meshfree finite element method, discontinuous Galerkin finite element method (DGFEM), Element-Free Galerkin Method (EFGM), Interpolating Element-Free Galerkin Method (IEFGM), etc.

Finite Difference Method

FDMs convert a linear ordinary differential equations (ODE) or non-linear partial differential equations (PDE) into a system of equations that can be solved by matrix algebra techniques. The reduction of the differential equation to a system of algebraic equations makes the problem of finding the solution to a given ODE/PDE ideally suited to modern computers, hence the widespread use of FDMs in modern numerical analysis. Today, FDMs are the dominant approach to numerical solutions of PDEs.

Derivation from Taylor's polynomial

First, assuming the function whose derivatives are to be approximated is properly-behaved, by Taylor's theorem, we can create a Taylor series expansion,

$$f(x_0+h)=f(x_0)+\frac{f'(x_0)}{1!}h+\frac{f^{(2)}(x_0)}{2!}h^2+\cdots+\frac{f^{(n)}(x_0)}{n!}h^n+R_n(x),$$

where $n!$ denotes the factorial of n, and $R_n(x)$ is a remainder term, denoting the difference between the Taylor polynomial of degree n and the original function. We will derive an approximation for the first derivative of the function "f" by first truncating the Taylor polynomial:

$$f(x_0+h)=f(x_0)+f'(x_0)h+R_1(x),$$

Setting, x_0=a we have,

$$f(a+h)=f(a)+f'(a)h+R_1(x),$$

Dividing across by h gives:

$$\frac{f(a+h)}{h}=\frac{f(a)}{h}+f'(a)+\frac{R_1(x)}{h}$$

Solving for f'(a):

$$f'(a)=\frac{f(a+h)-f(a)}{h}-\frac{R_1(x)}{h}$$

Assuming that $R_1(x)$ is sufficiently small, the approximation of the first derivative of "f" is:

$$f'(a)\approx\frac{f(a+h)-f(a)}{h}.$$

Accuracy and Order

The error in a method's solution is defined as the difference between the approximation and the exact analytical solution. The two sources of error in finite difference methods are round-off error, the loss of precision due to computer rounding of decimal quantities, and truncation error or discretization error, the difference between the exact solution of the original differential equation and the exact quantity assuming perfect arithmetic (that is, assuming no round-off).

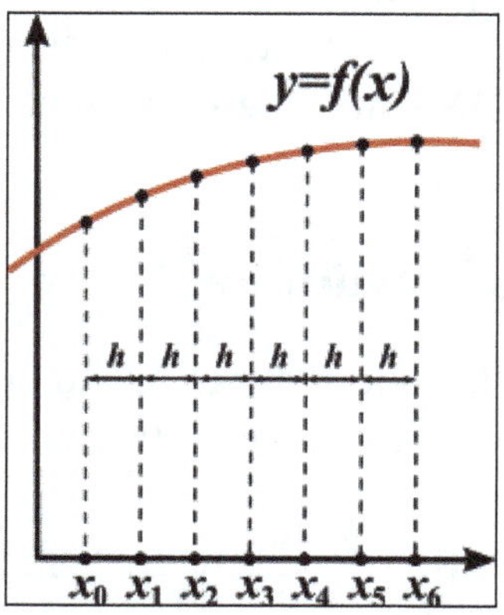

The finite difference method relies on discretizing a function on a grid.

To use a finite difference method to approximate the solution to a problem, one must first discretize the problem's domain. This is usually done by dividing the domain into a uniform grid. This means that finite-difference methods produce sets of discrete numerical approximations to the derivative, often in a "time-stepping" manner.

An expression of general interest is the local truncation error of a method. Typically expressed using Big-O notation, local truncation error refers to the error from a single application of a method. That is, it is the quantity $f'(x_i) - f_i'$ if $f'(x_i)$ refers to the exact value and f_i' to the numerical approximation. The remainder term of a Taylor polynomial is convenient for analyzing the local truncation error. Using the Lagrange form of the remainder from the Taylor polynomial for $f(x_o + h)$, which is,

$$R_n(x_o + h) = \frac{f^{(n+1)}(\xi)}{(n+1)!}(h)^{n+1}, \text{ where } x_o < \xi < x_o + h,$$

the dominant term of the local truncation error can be discovered. For example, again using the forward-difference formula for the first derivative, knowing that $f(x_i) = f(x_o + ih)$,

$$f(x_o + ih) = f(x_o) + f'(x_o)ih + \frac{f''(\xi)}{2!}(ih)^2,$$

and with some algebraic manipulation, this leads to,

$$\frac{f(x_o + ih) - f(x_o)}{ih} = f'(x_o) + \frac{f''(\xi)}{2!}ih,$$

and further noting that the quantity on the left is the approximation from the finite difference method and that the quantity on the right is the exact quantity of interest plus a remainder, clearly that remainder is the local truncation error. A final expression of this example and its order is:

$$\frac{f(x_o + ih) - f(x_o)}{ih} = f'(x_o) + O(h).$$

This means that, in this case, the local truncation error is proportional to the step sizes. The quality and duration of simulated FDM solution depends on the discretization equation selection and the step sizes (time and space steps). The data quality and simulation duration increase significantly with smaller step size. Therefore, a reasonable balance between data quality and simulation duration is necessary for practical usage. Large time steps are useful for increasing simulation speed in practice. However, time steps which are too large may create instabilities and affect the data quality.

The von Neumann and Courant-Friedrichs-Lewy criteria are often evaluated to determine the numerical model stability.

Ordinary Differential Equation

For example, consider the ordinary differential equation,

$$u'(x)=3u(x)+2.$$

The Euler method for solving this equation uses the finite difference quotient,

$$\frac{u(x+h)-u(x)}{h}\approx u'(x)$$

to approximate the differential equation by first substituting it for u'(x) then applying a little algebra (multiplying both sides by h, and then adding u(x) to both sides) to get,

$$u(x+h)=u(x)+h(3u(x)+2).$$

The last equation is a finite-difference equation, and solving this equation gives an approximate solution to the differential equation.

The Heat Equation

Consider the normalized heat equation in one dimension, with homogeneous Dirichlet boundary conditions,

$$U_t=U_{xx}$$

$$U(0,t)=U(1,t)=0 \text{ (boundary condition)}$$

$$U(x,0)=U_0(x) \text{ (initial condition)}$$

One way to numerically solve this equation is to approximate all the derivatives by finite differences. We partition the domain in space using a mesh $x_0,\ldots,x_J$ and in time using a mesh $t_0,\ldots,t_N$. We assume a uniform partition both in space and in time, so the difference between two consecutive space points will be *h* and between two consecutive time points will be *k*. The points,

$$u(x_j,t_n)=u_j^n$$

will represent the numerical approximation of $u(x_j,t_n)$.

Explicit Method

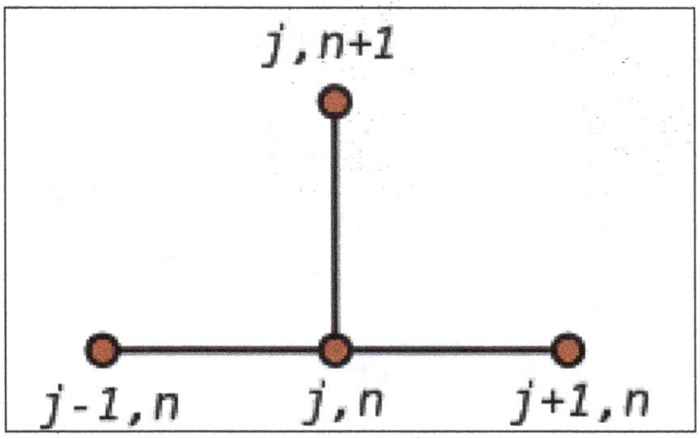

The stencil for the most common explicit method for the heat equation.

Using a forward difference at time t_n and a second-order central difference for the space derivative at position x_j (FTCS) we get the recurrence equation:

$$\frac{u_j^{n+1}-u_j^n}{k}=\frac{u_{j+1}^n-2u_j^n+u_{j-1}^n}{h^2}.$$

This is an explicit method for solving the one-dimensional heat equation.

We can obtain u_j^{n+1} from the other values this way:

$$u_j^{n+1}=(1-2r)u_j^n+ru_{j-1}^n+ru_{j+1}^n$$

where $r=k/h^2$.

So, with this recurrence relation, and knowing the values at time *n*, one can obtain the corresponding values at time *n*+1. u_0^n and u_J^n must be replaced by the boundary conditions, in this example they are both 0.

This explicit method is known to be numerically stable and convergent whenever $r \le 1/2$. The numerical errors are proportional to the time step and the square of the space step:

$$\Delta u=O(k)+O(h^2)$$

Implicit Method

If we use the backward difference at time t_{n+1} and a second-order central difference for the space derivative at position x_j (The Backward Time, Centered Space Method "BTCS") we get the recurrence equation:

$$\frac{u_j^{n+1}-u_j^n}{k}=\frac{u_{j+1}^{n+1}-2u_j^{n+1}+u_{j-1}^{n+1}}{h^2}.$$

This is an implicit method for solving the one-dimensional heat equation.

We can obtain u_j^{n+1} from solving a system of linear equations:

$$(1+2r)u_j^{n+1} - ru_{j-1}^{n+1} - ru_{j+1}^{n+1} = u_j^n$$

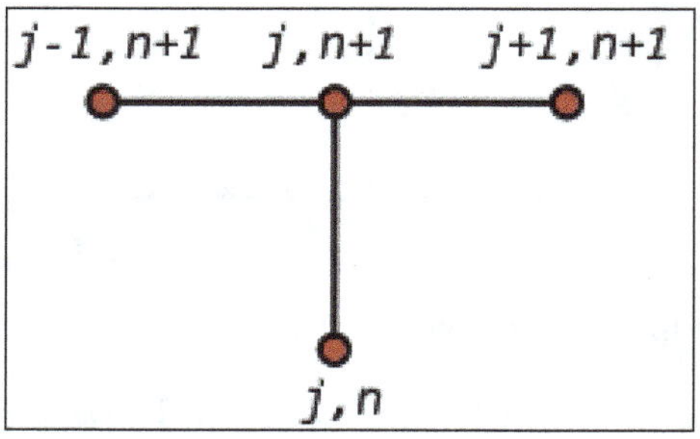

The implicit method stencil.

The scheme is always numerically stable and convergent but usually more numerically intensive than the explicit method as it requires solving a system of numerical equations on each time step. The errors are linear over the time step and quadratic over the space step:

$$\Delta u = O(k) + O(h^2).$$

Crank–Nicolson Method

Finally if we use the central difference at time $t_{n+1/2}$ and a second-order central difference for the space derivative at position x_j ("CTCS") we get the recurrence equation:

$$\frac{u_j^{n+1} - u_j^n}{k} = \frac{1}{2}\left(\frac{u_{j+1}^{n+1} - 2u_j^{n+1} + u_{j-1}^{n+1}}{h^2} + \frac{u_{j+1}^n - 2u_j^n + u_{j-1}^n}{h^2}\right).$$

This formula is known as the Crank–Nicolson method.

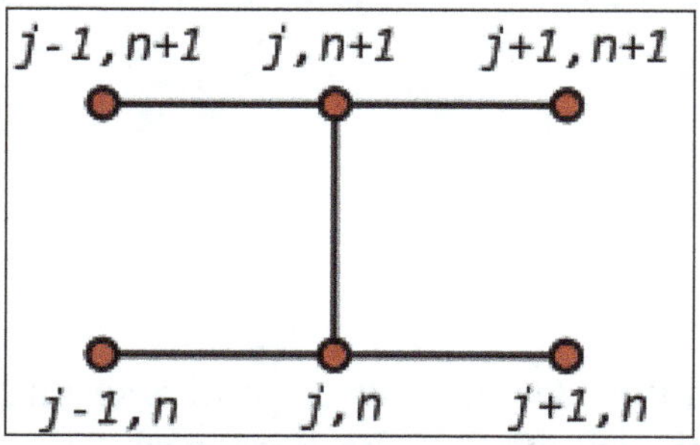

The Crank–Nicolson stencil.

We can obtain u_j^{n+1} from solving a system of linear equations:

$$(2+2r)u_j^{n+1} - ru_{j-1}^{n+1} - ru_{j+1}^{n+1} = (2-2r)u_j^n + ru_{j-1}^n + ru_{j+1}^n$$

The scheme is always numerically stable and convergent but usually more numerically intensive as it requires solving a system of numerical equations on each time step. The errors are quadratic over both the time step and the space step:

$$\Delta u = O\left(k^2\right) + O\left(h^2\right).$$

Usually the Crank–Nicolson scheme is the most accurate scheme for small time steps. The explicit scheme is the least accurate and can be unstable, but is also the easiest to implement and the least numerically intensive. The implicit scheme works the best for large time steps.

Comparison

The figures below present the solutions given by the above methods to approximate the heat equation,

$$U_t = \alpha U_{xx}, \quad \alpha = \frac{1}{\pi^2},$$

with the boundary condition,

$$U(0,t) = U(1,t) = 0.$$

The exact solution is

$$U(x,t) = \frac{1}{\pi^2} e^{-t} \sin(\pi x).$$

Comparison of Finite Difference Methods

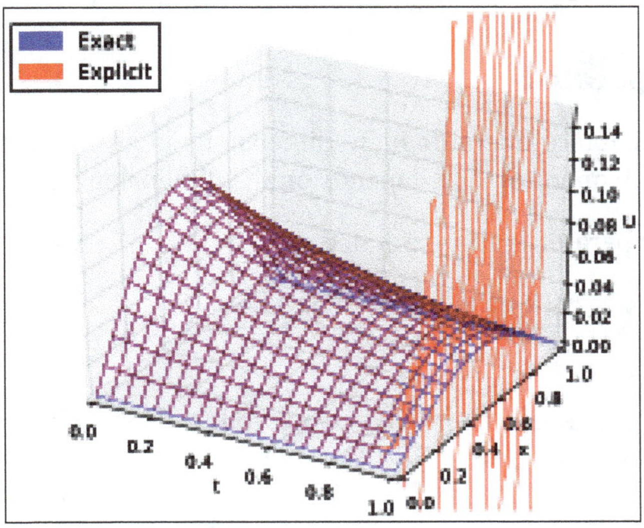

Explicit method (not stable).

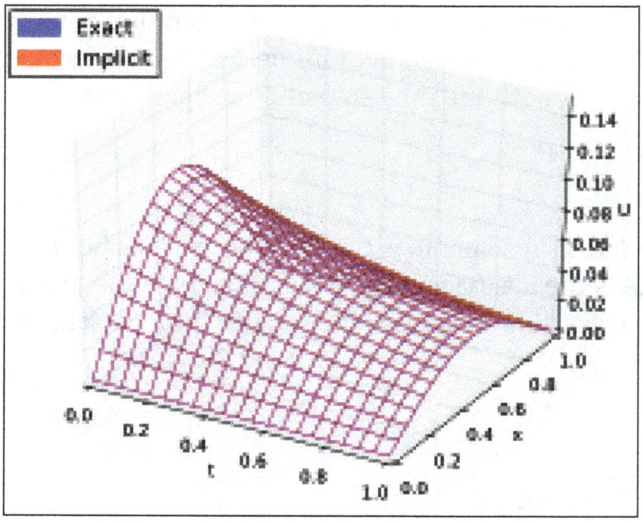

Implicit method (stable).

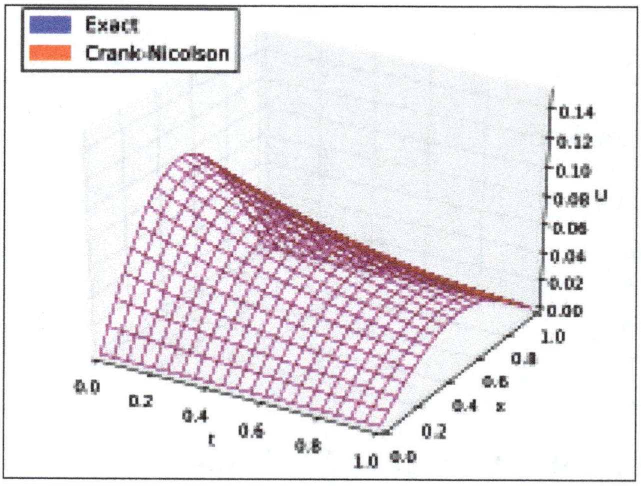

Crank-Nicolson method (stable).

The Laplace Operator

The (continuous) Laplace operator in n-dimensions is given by $\Delta u(x)=\sum_{i=1}^{n}\partial_i^2 u(x)$. The discrete Laplace operator $\Delta_h u$ depends on the dimension n.

In 1D the Laplace operator is approximated as,

$$\Delta u(x)=u''(x)\approx\frac{u(x-h)-2u(x)+u(x+h)}{h^2}=:\Delta_h u(x).$$

This approximation is usually expressed via the following stencil,

$$\frac{1}{h^2}\begin{bmatrix}1 & -2 & 1\end{bmatrix}$$

and which represents a symmetric, tridiagonal matrix. For an equidistant grid one gets a Toeplitz matrix.

The 2D case shows all the characteristics of the more general nD case. Each second partial derivative needs to be approximated similar to the 1D case,

$$\begin{aligned}\Delta u(x,y) &= u_{xx}(x,y) + u_{yy}(x,y)\\ &\approx \frac{u(x-h,y)-2u(x,y)+u(x+h,y)}{h^2} + \frac{u(x,y-h)-2u(x,y)+u(x,y+h)}{h^2}\\ &= \frac{u(x-h,y)+u(x+h,y)-4u(x,y)+u(x,y-h)+u(x,y+h)}{h^2}\\ &=: \Delta_h u(x,y),\end{aligned}$$

which is usually given by the following stencil:

$$\frac{1}{h^2}\begin{bmatrix} & 1 & \\ 1 & -4 & 1 \\ & 1 & \end{bmatrix}.$$

Consistency

Consistency of the above-mentioned approximation can be shown for highly regular functions, such as $u \in C^4(\Omega)$. The statement is,

$$\Delta u - \Delta_h u = \mathcal{O}(h^2).$$

To proof this one needs to substitute Taylor Series expansions up to order 3 into the discrete Laplace operator.

Properties

Subharmonic

Similar to continuous subharmonic functions one can define subharmonic functions for finite-difference approximations u_h,

$$-\Delta_h u_h \le 0.$$

Mean Value

One can define a general stencil of *positive type* via,

$$\begin{bmatrix} & \alpha_N & \\ \alpha_W & -\alpha_C & \alpha_E \\ & \alpha_S & \end{bmatrix}, \quad \alpha_i > 0, \quad \alpha_C = \sum_{i\in\{N,E,S,W\}} \alpha_i.$$

If u_h is (discrete) subharmonic then the following *mean value property* holds,

$$u_h(x_C) \leq \frac{\sum_{i \in \{N,E,S,W\}} \alpha_i u_h(x_i)}{\sum_{i \in \{N,E,S,W\}} \alpha_i},$$

where the approximation is evaluated on points of the grid, and the stencil is assumed to be of positive type.

A similar mean value property also holds for the continuous case.

Maximum Principle

For a (discrete) subharmonic function u_h the following holds,

$$\max_{\Omega_h} u_h \leq \max_{\partial\Omega_h} u_h,$$

where $\Omega_h, \partial\Omega_h$, are discretizations of the continuous domain Ω, respectively the boundary $\partial\Omega$.

A similar maximum principle also holds for the continuous case.

The SBP-SAT Method

The SBP-SAT method is a stable and accurate technique for discretizing and imposing boundary conditions of a well-posed partial differential equation using high order finite differences.The method is based on finite differences where the differentiation operators exhibit summation-by-parts properties. Typically, these operators consist of differentiation matrices with central difference stencils in the interior with carefully chosen one-sided boundary stencils designed to mimic integration-by-parts in the discrete setting. Using the SAT technique, the boundary conditions of the PDE are imposed weakly, where the boundary values are "pulled" towards the desired conditions rather than exactly fulfilled. If the tuning parameters (inherent to the SAT technique) are chosen properly, the resulting system of ODE's will exhibit similar energy behavior as the continuous PDE, i.e. the system has no non-physical energy growth. This guarantees stability if an integration scheme with a stability region that includes parts of the imaginary axis, such as the fourth order Runge-Kutta method, is used. This makes the SAT technique an attractive method of imposing boundary conditions for higher order finite difference methods, in contrast to for example the injection method, which typically will not be stable if high order differentiation operators are used.

Method of Lines

The method of lines (MOL, NMOL, NUMOL) is a technique for solving partial differential equations (PDEs) in which all but one dimension is discretized. MOL allows standard, general-purpose methods and software, developed for the numerical integration of ODEs and DAEs, to be used. Many integration routines have been developed over the years in many different programming languages, and some have been published as open source resources.

The method of lines most often refers to the construction or analysis of numerical methods for partial differential equations that proceeds by first discretizing the spatial derivatives only and leaving the time variable continuous. This leads to a system of ordinary differential equations to which a numerical method for initial value ordinary equations can be applied. The method of lines in this context dates back to at least the early 1960s.

Application to Elliptical Equations

MOL requires that the PDE problem is well-posed as an initial value (Cauchy) problem in at least one dimension, because ODE and DAE integrators are initial value problem (IVP) solvers. Thus it cannot be used directly on purely elliptic partial differential equations, such as Laplace's equation. However, MOL has been used to solve Laplace's equation by using the *method of false transients*. In this method, a time derivative of the dependent variable is added to Laplace's equation. Finite differences are then used to approximate the spatial derivatives, and the resulting system of equations is solved by MOL. It is also possible to solve elliptical problems by a *semi-analytical method of lines*. In this method, the discretization process results in a set of ODE's that are solved by exploiting properties of the associated exponential matrix.

Recently, to overcome the stability issues associated with the method of false transients, a perturbation approach was proposed which was found to be more robust than standard method of false transients for a wide range of elliptic PDEs.

Finite Volume Method

The finite volume method (FVM) is a method for representing and evaluating partial differential equations in the form of algebraic equations. In the finite volume method, volume integrals in a partial differential equation that contain a divergence term are converted to surface integrals, using the divergence theorem. These terms are then evaluated as fluxes at the surfaces of each finite volume. Because the flux entering a given volume is identical to that leaving the adjacent volume, these methods are conservative. Another advantage of the finite volume method is that it is easily formulated to allow for unstructured meshes. The method is used in many computational fluid dynamics packages. "Finite volume" refers to the small volume surrounding each node point on a mesh.

Alternative to the finite difference method or finite element method, values are calculated at discrete places on a meshed geometry.

Example:

Consider a simple 1D advection problem:

$$\frac{\partial \rho}{\partial t}+\frac{\partial f}{\partial x}=0, \quad t\geq 0.$$

Here, $\rho=\rho(x,t)$ represents the state variable and $f=f(\rho(x,t))$ represents the flux or flow of ρ. Conventionally, positive f represents flow to the right while negative f represents flow to the left. If we assume that equation $\frac{\partial \rho}{\partial t}+\frac{\partial f}{\partial x}=0, \quad t\geq 0$ represents a flowing medium of constant area, we can sub-divide the spatial domain, x, into *finite* volumes or cells with cell centres indexed as i. For a particular cell, i, we can define the *volume average* value of $\rho_i(t)=\rho(x,t)$ at time $t=t_1$ and $x\in\left[x_{i-\frac{1}{2}},x_{i+\frac{1}{2}}\right]$, as,

$$\bar{\rho}_i(t_1)=\frac{1}{x_{i+\frac{1}{2}}-x_{i-\frac{1}{2}}}\int_{x_{i-\frac{1}{2}}}^{x_{i+\frac{1}{2}}}\rho(x,t_1)dx,$$

and at time $t=t_2$ as,

$$\bar{\rho}_i(t_2)=\frac{1}{x_{i+\frac{1}{2}}-x_{i-\frac{1}{2}}}\int_{x_{i-\frac{1}{2}}}^{x_{i+\frac{1}{2}}}\rho(x,t_2)dx,$$

where $x_{i-\frac{1}{2}}$ and $x_{i+\frac{1}{2}}$ represent locations of the upstream and downstream faces or edges respectively of the ith cell.

Integrating equation $\frac{\partial \rho}{\partial t}+\frac{\partial f}{\partial x}=0, \quad t\geq 0$ in time, we have:

$$\rho(x,t_2)=\rho(x,t_1)-\int_{t_1}^{t_2}f_x(x,t)dt,$$

where $f_x=\frac{\partial f}{\partial x}$.

To obtain the volume average of $\rho(x,t)$ at time $t=t_2$, we integrate $\rho(x,t_2)$ over the cell volume, $\left[x_{i-\frac{1}{2}},x_{i+\frac{1}{2}}\right]$ and divide the result by $\Delta x_i=x_{i+\frac{1}{2}}-x_{i-\frac{1}{2}}$, i.e.

$$\bar{\rho}_i(t_2)=\frac{1}{\Delta x_i}\int_{x_{i-\frac{1}{2}}}^{x_{i+\frac{1}{2}}}\left\{\rho(x,t_1)-\int_{t_1}^{t_2}f_x(x,t)dt\right\}dx.$$

We assume that f is well behaved and that we can reverse the order of integration. Also, recall that flow is normal to the unit area of the cell. Now, since in one dimension $f_x \triangleq \nabla \cdot f$, we can apply the divergence theorem, i.e. $\oint_v \nabla \cdot f dv = \oint_S f dS$, and substitute for the volume integral of the divergence with the values of $f(x)$ evaluated at the cell surface (edges $x_{i-\frac{1}{2}}$ and $x_{i+\frac{1}{2}}$) of the finite volume as follows:

$$\bar{\rho}_i(t_2) = \bar{\rho}_i(t_1) - \frac{1}{\Delta x_i}\left(\int_{t_1}^{t_2} f_{i+\frac{1}{2}} dt - \int_{t_1}^{t_2} f_{i-\frac{1}{2}} dt\right).$$

where $f_{i\pm\frac{1}{2}} = f\left(x_{i\pm\frac{1}{2}}, t\right)$.

We can therefore derive a *semi-discrete* numerical scheme for the above problem with cell centres indexed as i, and with cell edge fluxes indexed as $i \pm \frac{1}{2}$, by differentiating equation $\bar{\rho}_i(t_2) = \bar{\rho}_i(t_1) - \frac{1}{\Delta x_i}\left(\int_{t_1}^{t_2} f_{i+\frac{1}{2}} dt - \int_{t_1}^{t_2} f_{i-\frac{1}{2}} dt\right)$ with respect to time to obtain:

$$\frac{d\bar{\rho}_i}{dt} + \frac{1}{\Delta x_i}\left[f_{i+\frac{1}{2}} - f_{i-\frac{1}{2}}\right] = 0,$$

where values for the edge fluxes, $f_{i\pm\frac{1}{2}}$, can be reconstructed by interpolation or extrapolation of the cell averages. Equation $\frac{d\bar{\rho}_i}{dt} + \frac{1}{\Delta x_i}\left[f_{i+\frac{1}{2}} - f_{i-\frac{1}{2}}\right] = 0$, is *exact* for the volume averages; i.e., no approximations have been made during its derivation.

This method can also be applied to a 2D situation by considering the north and south faces along with the east and west faces around a node.

General Conservation Law

We can also consider the general conservation law problem, represented by the following PDE,

$$\frac{\partial \mathbf{u}}{\partial t} + \nabla \cdot \mathbf{f}(\mathbf{u}) = \mathbf{0}.$$

Here, $\mathbf{u}$ represents a vector of states and $\mathbf{f}$ represents the corresponding flux tensor. Again we can sub-divide the spatial domain into finite volumes or cells. For a particular cell, i, we take the volume integral over the total volume of the cell, v_i, which gives,

$$\int_{v_i} \frac{\partial \mathbf{u}}{\partial t} dv + \int_{v_i} \nabla \cdot \mathbf{f}(\mathbf{u}) dv = \mathbf{0}.$$

On integrating the first term to get the *volume average* and applying the *divergence theorem* to the second, this yields,

$$v_i \frac{d\bar{\mathbf{u}}_i}{dt} + \oint_{S_i} \mathbf{f}(\mathbf{u}) \cdot \mathbf{n}\, dS = \mathbf{o},$$

where S_i represents the total surface area of the cell and $\mathbf{n}$ is a unit vector normal to the surface and pointing outward. So, finally, we are able to present the general result equivalent to equation $\frac{\partial \mathbf{u}}{\partial t} + \nabla \cdot \mathbf{f}(\mathbf{u}) = \mathbf{o}$, i.e.

$$\frac{\overline{}}{dt} + \frac{}{v} \oint \mathbf{f}(\mathbf{u}) \cdot \mathbf{n}\, dS = \mathbf{o}.$$

Again, values for the edge fluxes can be reconstructed by interpolation or extrapolation of the cell averages. The actual numerical scheme will depend upon problem geometry and mesh construction. MUSCL reconstruction is often used in high resolution schemes where shocks or discontinuities are present in the solution.

Finite volume schemes are conservative as cell averages change through the edge fluxes.

Domain Decomposition Methods

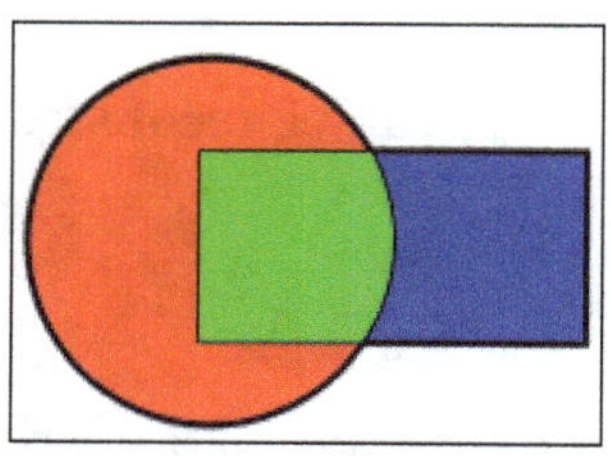

Domain decomposition methods.

In mathematics, numerical analysis, and numerical partial differential equations, domain decomposition methods solve a boundary value problem by splitting it into smaller boundary value problems on subdomains and iterating to coordinate the solution between adjacent subdomains. A coarse problem with one or few unknowns per subdomain is used to further coordinate the solution between the subdomains globally. The problems on the subdomains are independent, which makes domain decomposition methods suitable for parallel computing. Domain decomposition methods are typically used as preconditioners for Krylov space iterative methods, such as the conjugate gradient method or GMRES.

In overlapping domain decomposition methods, the subdomains overlap by more than the interface. Overlapping domain decomposition methods include the Schwarz alternating method and the additive Schwarz method. Many domain decomposition methods can be written and analyzed as a special case of the abstract additive Schwarz method.

In non-overlapping methods, the subdomains intersect only on their interface. In primal methods, such as Balancing domain decomposition and BDDC, the continuity of the solution across subdomain interface is enforced by representing the value of the solution on all neighboring subdomains by the same unknown. In dual methods, such as FETI, the continuity of the solution across the subdomain interface is enforced by Lagrange multipliers. The FETI-DP method is hybrid between a dual and a primal method.

Non-overlapping domain decomposition methods are also called iterative substructuring methods.

Mortar methods are discretization methods for partial differential equations, which use separate discretization on nonoverlapping subdomains. The meshes on the subdomains do not match on the interface, and the equality of the solution is enforced by Lagrange multipliers, judiciously chosen to preserve the accuracy of the solution. In the engineering practice in the finite element method, continuity of solutions between non-matching subdomains is implemented by multiple-point constraints.

Finite element simulations of moderate size models require solving linear systems with millions of unknowns. Several hours per time step is an average sequential run time, therefore, parallel computing is a necessity. Domain decomposition methods embody large potential for a parallelization of the finite element methods, and serve a basis for distributed, parallel computations.

1D Linear BVP

$$u''(x) - u(x) = 0$$

$$u(0) = 0, u(1) = 1$$

The exact solution is:

$$u(x) = \frac{e^{x} - e^{-x}}{e^{1} - e^{-1}}$$

Subdivide the domain into two subdomains, one from $\left[0, \frac{1}{2}\right]$ and another from $\left[\frac{1}{2}, 1\right]$.

In each of these two subdomains define interpolating functions $v_1(x)$ and $v_2(x)$ At the interface between these two subdomains the following interface conditions shall be imposed:

$$v_1\left(\frac{1}{2}\right) = v_2\left(\frac{1}{2}\right)$$

$$v_{1'}\left(\frac{1}{2}\right) = v_{2'}\left(\frac{1}{2}\right)$$

Let the interpolating functions be defined as:

$$v_1(x)=\sum_{n=0}^{N}u_n T_n\left(y_1(x)\right)$$

$$v_2(x)=\sum_{n=0}^{N}u_{n+N} T_n\left(y_2(x)\right)$$

$$y_1(x)=4x-1$$

$$y_2(x)=4x-3$$

Where $T_n(y)$ is the nth cardinal function of the chebyshev polynomials of the first kind with input argument y.

If N=4 then the following approximation is obtained by this scheme:

$$u_1=0.06236$$

$$u_2=0.21495$$

$$u_3=0.37428$$

$$u_4=0.44341$$

$$u_5=0.51492$$

$$u_6=0.69972$$

$$u_7=0.90645$$

This was obtained with the following MATLAB code.

```
clear all

N=4;

a1=0; b1=1/2;


[T D1 D2 E1 E2 x xsub]=cheb(N,a1,b1); % the diff matrices on [0,1/2]
are the same

%as those on [1/2 1].

I=eye(N+1);
```

```
H=D2-I;
H1=[[1 zeros(1,N)]; H(2:end-1,:); [zeros(1,N) 1]];
H1=[H1 [zeros(N,N+1); -[1 zeros(1,N)]]];
H2=[D1(1,:); H(2:end-1,:); [zeros(1,N) 1]];
H2=[[-D1(N+1,:); zeros(N,N+1)] H2];
K=[H1; H2];
F=[zeros(2*N+1,1); 1];
u=K\F;
xx=-cos(pi*(0:N)'/N);
x1=1/4*(xx+1); x2=1/4*(xx+3);
x=[x1; x2];
uex=(exp(x)-exp(-x))./(exp(1)-exp(-1));
```

Finite Volume Method

The Finite Volume Method (FVM) is a discretization method for the approximation of a single or a system of partial differential equations expressing the conservation, or balance, of one or more quantities. These partial differential equations (PDEs) are often called conservation laws; they may be of different nature, e.g. elliptic, parabolic or hyperbolic, and they are used as models in a wide number of fields, including physics, biophysics, chemistry, image processing, finance, dynamic reliability. They describe the relations between partial derivatives of unknown fields such as temperature, concentration, pressure, molar fraction, density of electrons or probability density function, with respect to variables within the domain (space, time) under consideration.

As in the finite element method, a mesh is constructed, which consists in a partition of the domain where the space variable lives. The elements of the mesh are called control volumes. The integration of the PDE over each control volume results in a balance equation. The set of balance equations is then discretized with respect to a set of discrete unknowns. The main issue is the discretization of the fluxes at the boundaries of each control volume: in order for the FVM to be efficient, the numerical fluxes are generally,

- Conservative, i.e. the flux entering a control volume from its neighbour must be the opposite of the one entering the neighbour from the control volume.
- Consistent, i.e. the numerical flux of a regular function interpolation tends to the continuous flux as the mesh size vanishes.

It is sometimes possible to discretize the fluxes at the boundaries of the control volume by the finite difference method (FDM). In this case, the method has often been referred to as a finite difference method or conservative finite difference method. The specificity of the FVM with respect to the FDM is that the discretization is performed on the local balance equations, rather than the PDE: the fluxes on boundaries of the control volumes are discretized, rather than the continuous differential operator.

The resulting system of discrete equations depends on a discrete (finite) set of unknowns, and may be either linear or non linear, depending on the original problem itself; this system is then solved exactly or approximately, using for example direct or iterative solvers in the case of linear equations and fixed point or Newton type methods in the case of nonlinear equations.

Let us emphasize that there is no systematic way to derive the discrete system from the continuous one; in fact, the physical or engineering knowledge of the problem is in many cases the key ingredient in order to obtain a scheme which satisfies the expected properties of accuracy, robustness, and computational cost. Let us mention that in some cases, the engineering problem is initially set at a discrete level: the model consists of a set of discrete balance equations and therefore written as a finite volume scheme, and the continuous model of the problem is not explicitly given.

Many of these issues have often been addressed in computational fluid mechanics.

Fundamental Principles

Consider the following PDE under conservative form,

$$\partial_t A(x,t) + \nabla \cdot F(x,t) = S(x,t)$$

where the space variable x belongs to the domain $\Omega \subset \mathbb{R}^d$ (d is the space dimension, greater or equal to 1), and the time variable t belongs to sometime interval $[0,T]$, with $T > 0$. The scalar function A, defined in $\Omega \times [0,T]$, expresses the density of some quantity, and $\partial_t A$ denotes its time derivative. The function F, defined in $\Omega \times [0,T]$ and valued in $\mathbb{R}^d$, expresses the flux of this quantity, and $\nabla \cdot F$ denotes its space divergence. The function S, defined in $\Omega \times [0,T]$ and valued in $\mathbb{R}$, denotes some source term. Some initial condition $A(x,0) = A_{ini}(x)$ for $x \in \Omega$ is imposed, where the function A_{ini} is defined in Ω and valued in $\mathbb{R}$, as well as some boundary conditions, which depend on the considered equation. The functions A and F are not necessarily regular, so that the derivatives involved in equation $\partial_t A(x,t) + \nabla \cdot F(x,t) = S(x,t)$ may be weak derivatives.

These functions $A,F, |S$ are assumed to be related to a set of unknown fields $(u_j)_{j=1,\ldots,}$ N, where u_j is an unknown function defined from $\Omega \times [0,T]$ to $\mathbb{R}$. Such relationships may for instance be written as,

$$A(x,t) = A(u_1(x,t),\ldots,u_N(x,t),x,t),$$
$$S(x,t) = S(u_1(x,t),\ldots,u_N(x,t),x,t),$$
$$F(x,t) = F(u_1(x,t),\ldots,u_N(x,t),\nabla u_1(x,t),\ldots,\nabla u_N(x,t),x,t),$$

where the functions A, S and F are given. Classical examples with $N = 1$ we then denote by u the unique component u_1), are:

- The heat equation which corresponds to $A(x,t) = u(x,t)$, and u s the temperature; $F(x,t) = -\Lambda(x)\nabla u(x,t)$ (where for any $x \in R^d$, $\Lambda(x)$ is a linear operator from $\mathbb{R}^d$ to $\mathbb{R}^d$), $S(x,t) = S(x,t)$.
- The Poisson equation, which is the steady version of the heat equations $A(x,t) = 0$,$F(x,t) = -\Lambda(x)\nabla u(x)$,$S(x,t) = S(x)$.
- The linear convection equation $A(x,t) = u(x,t)$,$F(x,t) = u(x,t)V(x,t)$ where V is a function defined from $\overline{\Omega}\times[0,T]$ to R^d, u is, say, a concentration, and V its transport velocity.

The domain Ω is partitioned into a mesh, denoted by $\mathcal{M}$. In general, the mesh $\mathcal{M}$ is a finite family of nonoverlapping subsets of Ω, which satisfy a number of regularity properties; some of these properties are needed for the definition of the scheme, while others are required to establish the precision and the convergence of the method, depending on the type of term that needs to be discretized. The elements of $\mathcal{M}$ denoted by K, L are called the control volumes; the measure of a control volume K (its length if $d = 1$, area if $d = 2$, volume if $d = 3$) is denoted by $|K|$. The boundary ∂K of each control volume K is partitioned into the finite set ε_K, called the set of the faces $(d = 3)$ or edges $(d = 2)$ or extremities $(d = 1)$ of K. An element σ of ε_K is either located on the boundary of Ω, or belongs to $\varepsilon_K \cap \varepsilon_L$ where K and L are two adjacent control volumes.

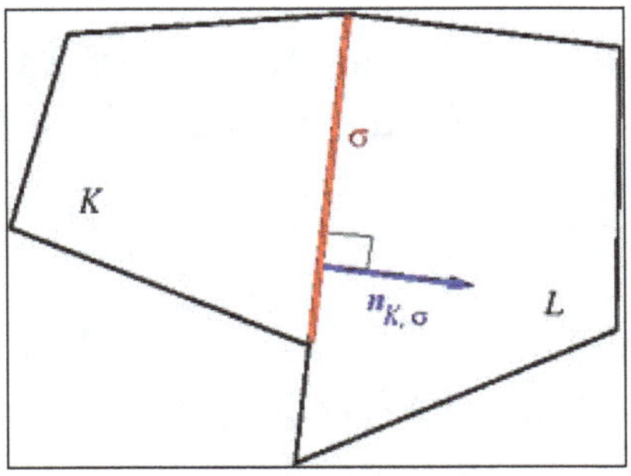

Two control volumes with a common face.

The $(d-1)$-dimensional measure of a face $\sigma \in \varepsilon_K$ (its length if $d = 2$, area if $d = 3$) is denoted by $|\sigma|$. A strictly increasing finite sequence $t^{(0)} = 0 < t^{(1)} < \ldots < t^{(N)} = T$ is defined for the time discretization, and one denotes $\delta t^{(n)} = t^{(n)} - t^{(n-1)}$, for $n = 1, \ldots, N$. The classical finite volume approximation of equation $\partial_t A(x,t) + \nabla \cdot F(x,t) = S(x,t)$ relies on the approximation of the balance equations on the control volumes between time $t^{(n-1)}$ and $t^{(n)}$ (in fact, the balance equations usually precede the continuous equation $\partial_t A(x,t) + \nabla \cdot F(x,t) = S(x,t)$ in the derivation of the model; this is the reason why, in several engineering applications, the FVM is applied directly to the discrete balance equations without any knowledge of the continuous equations). These balance equations may be obtained by integrating equation $\partial_t A(x,t)+\nabla\cdot F(x,t)=S(x,t)$ on $K\times\left[t^{(n-1)},t^{(n)}\right]$ and using the divergence formula,

$$\int_K\left(A\left(x,t^{(n)}\right)-A\left(x,t^{(n-1)}\right)\right)dx+\sum_{\sigma\in\varepsilon_K}\int_{t^{(n-1)}}^{t^{(n)}}\int_\sigma F(x,t)\cdot n_{K,\sigma}(x)ds(x)dt=\int_{t^{(n-1)}}^{t^{(n)}}\int_K S(x,t)dxdt,$$

where ds(x) denotes the integration with respect to the (d−1)-dimensional measure of the boundary and $n_{K,\sigma}(x)$ the unit vector normal to σ at point x, outward to K. The FV scheme is a discretization of the above balance equation, and may be written as,

$$|K|(A_K^{(n)}-A_K^{(n-1)})+\delta t^{(n)}\sum_{\sigma\in\varepsilon_K}|\sigma|F_{K,\sigma}^{(n)}=\delta t^{(n)}|K|S_K^{(n)},$$

where,

$$A_K^{(n)}(\text{resp.}A_K^{(0)})\text{ is an approximation of }\frac{1}{|K|}\int_K A(x,t^{(n)})dx\ (\text{resp.}\frac{1}{|K|}\int_K A_{ini}(x)dx),$$

$$F_{K,\sigma}^{(n)}\text{ is an approximation of }\phi_{K,\sigma}^{(n)}=\frac{1}{\delta t^{(n)}|\sigma|}\int_{t^{(n-1)}}^{t^{(n)}}\int_\sigma F(x,t)\cdot n_{K,\sigma}(x)ds(x)dt,$$

and

$$S_K^{(n)}\text{ is an approximation of }\frac{1}{\delta t^{(n)}|K|}\int_{t^{(n-1)}}^{t^{(n)}}\int_K S(x,t)dxdt.$$

For a face σ common to two control volumes K and L, since $n_{K,\sigma}(x)+n_{L,\sigma}(x)=0$ for $x\in\sigma$, the normal fluxes in the original problem satisfy the conservation property $\phi_{K,\sigma}^{(n)}+\phi_{L,\sigma}^{(n)}=0$. A discrete version of this property of conservation is then required,

$$F_{K,\sigma}^{(n)}+F_{L,\sigma}^{(n)}=0.$$

This relation is one of the pillars of the classical FVM; moreover, it is crucial in the mathematical analysis for proving convergence properties.

Discrete expressions in terms of discrete unknowns must then be provided for the terms $A_K^{(n)}$, $F_{K,\sigma}^{(n)}$ and $S_K^{(n)}$. These discrete unknowns are usually expected to be approximations of the unknown functions $(u_j)_{j=1,\ldots,N}$ at different locations (grid blocks, faces, vertices) and times: hence $u_{j,K}^{(n)}$ (resp. $u_{j,\sigma}^{(n)}|$) denotes the approximation of function u_j in control volume K (resp. at face σ) at time $t^{(n)}$. For example, considering the case given in equation $A(x,t)\ldots,\nabla u_N(x,t),x,t)$, it is generally possible to simply use the function A to define $A_K^{(n)}$, using an expression such as,

$$A_K^{(n)}=\frac{1}{|K|}\int_K A(u_{1,K}^{(n)},\ldots,u_{N,K}^{(n)},x,t^{(n)})dx.$$

It is generally less simple to derive from $A(x,t)\ldots,\nabla u_N(x,t),x,t)$, an expression for $F_{K,\sigma}^{(n)}$ with respect to the discrete variables, since there is no systematic method to use the function F which ensures accuracy and stability on general grids.

The arguments for the discrete expression of $S_K^{(n)}$ are generally $u_{j,K}^{(m)},\ldots, m = n$ or $n-1$.

Discretization of Diffusion Fluxes

We consider here a diffusive flux $F(x,t)$ of the form $F(x,t) = -\Lambda(x,t)\nabla u(x,t)$, where $\Lambda(x,t)$ is a linear operator from $\mathbb{R}^d$ to $\mathbb{R}^d$, and u is one of the continuous unknowns $(u_j)_{j=1,\ldots,N}$ of the problem. Such an expression of the flux is encountered when using e.g. Fourier's law (heat conduction), Darcy's law (flow in porous media) or Fick's law (chemical diffusion).

In some cases, a simple, conservative and consistent discrete expression for $F_{K,\sigma}^{(n)}$ may be obtained. This is the case when $\Lambda(x,t)$ is the identity operator and when the grid satisfies the following "orthogonal property": in each control volume $K \in M$, there exists a particular point x_K, called the "center" (or centroid) of the control volume, such that, for any adjacent control volume L to K with common face σ, the straight line (x_K, x_L) is orthogonal to σ.

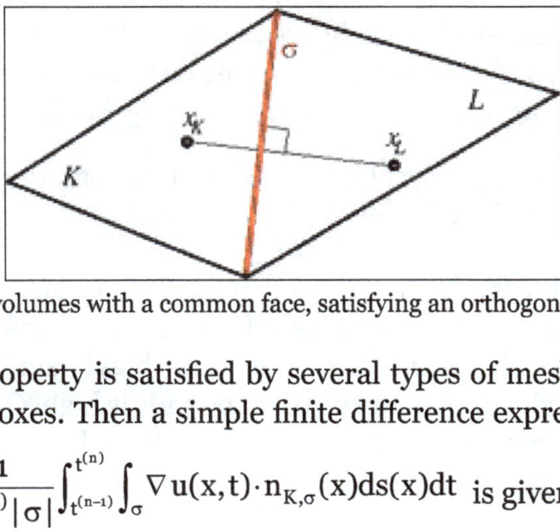

Two control volumes with a common face, satisfying an orthogonality property.

This orthogonality property is satisfied by several types of meshes, such as triangles, rectangles, Voronoï boxes. Then a simple finite difference expression for the approximation of $\phi_{K,\sigma}^{(n)} = -\frac{1}{\delta t^{(n)}|\sigma|}\int_{t^{(n-1)}}^{t^{(n)}}\int_\sigma \nabla u(x,t)\cdot n_{K,\sigma}(x)ds(x)dt$ is given by,

$$F_{K,\sigma}^{(n)} = \frac{u_K^{(m)} - u_L^{(m)}}{d(x_K, x_L)},$$

where $d(x_K, x_L)$ denotes the Euclidean distance between the points x_K and x_L, and provides an implicit (resp. explicit) approximation if $m = n$ (resp. $m = n-1$). This is the famous "two-point flux approximation" scheme, which has been widely used. It is simple, and also much appreciated because it respects a discrete version of the maximum principle; in particular, the approximate solution does not oscillate and stays within its physical bounds. The error analysis relies on the conservativity and consistency of the numerical flux.

In the general case of unstructured meshes and general (possibly anisotropic and heterogeneous) linear operators $\Lambda(x,t)$, a two point flux formula is not consistent in general, and yields an approximation which might not converge to the appropriate solution.

There are several ways to derive expressions for $F_{K,\sigma}^{(n)}$ in the general case (intensive research is going on for the analysis of these schemes and their implementation):

- The flux may for instance be approximated by a consistent formula using several points and respecting the conservativity property, such as in the MPFA (multiple point flux approximation) scheme.
- Expressions for $F_{K,\sigma}^{(n)}$ may also be obtained thanks to an identification procedure, using a variational expression and discrete gradients. Such a technique is used in mixed finite volume, hybrid finite volumes (and the SUSHI scheme) and the mimetic method. These three methods can in fact be viewed as elements of one unified family of schemes.
- Another way to express $F_{K,\sigma}^{(n)}$ in the anisotropic case is to partition the domain Ω in more than one mesh (typically two in 2D and three in 3D), and use the differences of the unknowns in the different meshes to reconstruct an expression for $F_{K,\sigma}^{(n)}$. This is the principle of the discrete duality FVM.
- The expression of $F_{K,\sigma}^{(n)}$ may also be inspired by the finite element framework, as in the example of Finite Element-Finite Volume methods.

Unfortunately, all the linear schemes devised in the general case of unstructured meshes and general linear operators do not seem to satisfy a discrete version of the maximum principle; for severely distorted meshes or for highly anisotropic operators, the risk of oscillations and violation of the physical bounds is high. Ongoing research using in particular nonlinear schemes for linear problems is under way to avoid these drawbacks.

Approximation of Convection Terms

Nonlinear Scalar Convection Equations

Consider the partial differential equation,

$$\partial_t u(x,t) + \nabla \cdot F(u(x,t),x,t) = 0.$$

This is the conservation equation $\partial_t A(x,t) + \nabla \cdot F(x,t) = S(x,t)$ with $A(x,t) = u(x,t)$,$F(x,t) = F(u(x,t),x,t)$ and $S(x,t) = 0$. The linear transport equation is a particular case, with $F(u,x,t)=uV(x,t)$, and V is defined from $\Omega\times[0,T]$ to $\mathbb{R}^d$. The discrete unknowns are the values $u_K^{(n)}$, which are expected to be approximations of u in the control volumes $K\in M|$ at time $t^{(n)}$. Since the convection flux is of order 0 (no derivative involved) its approximation does not require any assumption on the mesh, contrary to the diffusion flux. Nevertheless, the fluxes must be carefully approximated in order to ensure the stability (and convergence) of the scheme. Consider the flux $\phi_{K,\sigma}^{(n)}$ through an interface σ at time $t^{(n)}$, seen as a function of the solution,

$$u:\phi_{K,\sigma}^{(n)}(u)=\frac{1}{\delta t^{(n)}\,|\sigma|}\int_{t^{(n-1)}}^{t^{(n)}}\int_{\sigma}F(u(x,t),x,t)\cdot n_{K,\sigma}(x)ds(x)dt.$$

A numerical monotone flux associated to $\phi_{K,\sigma}^{(n)}(u)$ is a function $G_{K,\sigma}^{(n)}$ from $\mathbb{R}^2$ to $\mathbb{R}$ such that:

- $G_{K,\sigma}^{(n)}(a,b)=-G_{L,\sigma}^{(n)}(b,a)$, for an interface $\sigma\in\varepsilon_K\cap\varepsilon_L$.
- $G_{K,\sigma}^{(n)}$ is non-decreasing with respect to its first variable and non-increasing with respect to its second variable.
- $G_{K,\sigma}^{(n)}$ is locally Lipschitz-continuous with respect to its variables.
- $G_{K,\sigma}^{(n)}(a,a)=\phi_{K,\sigma}^{(n)}(a)$, for any a respecting the bounds of the solution.

Then FV schemes are obtained by approximating $\phi_{K,\sigma}^{(n)}(u)$ by,

- $F_{K,\sigma}^{(n)}=G_{K,\sigma}^{(n)}\left(u_K^{(m)},u_L^{(m)}\right)$, for an interface $\sigma\in\varepsilon_K\cap\varepsilon_L$.
- $F_{K,\sigma}^{(n)}=G_{K,\sigma}^{(n)}\left(u_K^{(m)},u_\sigma^{(n)}\right)$, for an interface $\sigma\subset\partial\Omega$, where $u_\sigma^{(n)}$ is a suitable approximation of the boundary condition on $\sigma\times]t^{(n-1)},t^{(n)}[$, with $m=n-1$ (explicit scheme) or $m=n$ (implicit scheme).

Provided that, for an explicit scheme, the so-called CFL condition be satisfied (this condition prescribes a bound for the time step with respect to the size of the mesh), these schemes satisfy the following properties:

- They are L^∞-stable, in the sense that the bounds of the initial and boundary data are preserved.
- They are Total Variation Diminishing (TVD), in the case of structured meshes.

Hyperbolic Systems

We now consider a system of N one-dimensional conservation laws $\partial_t A_j(x,t)+\partial_x F_j(x,t)=0$, where, for $j=1,\dots,N, A_j(x,t)=u_j(x,t)$ and the scalar flux $F_j(x,t)$ is given by $F_j(x,t)=F_j(u_1(x,t),u_2(x,t),\dots,u_N(x,t),x,t)$. The continuous problem itself is not easy to study from a mathematical point of view, and open questions include existence and uniqueness of a physically relevant solution in the general system case. The continuous fluxes $(\phi_j)_{K,\sigma}^{(n)}$ between times $t^{(n-1)}$ and $t^{(n)}$ through an interface σ (which is a real point in the one-dimensional framework) read

$$(\phi_j)_{K,\sigma}^{(n)} = \frac{n_{K,\sigma}}{\delta t^{(n)}} \int_{t^{(n-1)}}^{t(n)} F_j(u_1(\sigma,t),\ldots,u_N(\sigma,t),\sigma,t)dt, j=1,\ldots,N,$$

with $n_{K,\sigma}=1$ or $n_{K,\sigma}=-1$, depending on the position of σ with respect to K. A time-explicit FVM approximation reads,

$$|K|\left(u_{j,K}^{(n)} - u_{j,K}^{(n-1)}\right) + \delta t^{(n)} \sum_{\sigma\in\varepsilon_K} \left(F_j\right)_{K,\sigma}^{(n)} = 0, j=1,\ldots,N,$$

Where $\left(F_j\right)_{K,\sigma}^{(n)}$ is an approximation of $(\phi_j)_{K,\sigma}^{(n)}$. Such an approximation may be performed by the very famous Godunov scheme. The idea is, for given $K,\sigma,n,$ to approximate the flux $(\phi_j)_{K,\sigma}^{(n)}$ by $\left(F_j\right)_{K,\sigma}^{(n)} = f_j\left(w_1^R,\ldots,w_N^R\right)$, where f_j is the function defined from $\mathbb{R}^N$ to $\mathbb{R}$ by,

$$f_j(v_1,\ldots,v_N) = \frac{n_{K,\sigma}}{\delta t^{(n)}} \int_{t^{(n-1)}}^{t^{(n)}} F_j(v_1,\ldots,v_N,\sigma,t)dt, j=1,\ldots,N, \forall (v_1,\ldots,v_N) \in \mathbb{R}^N,$$

and $\left(w_1^R,\ldots,w_N^R\right)$ is solution at $\xi=0$ for all $\tau>0$ of the so-called one-dimensional Riemann problem, defined by,

$$\begin{aligned}
&\partial_\tau w_j(\xi,\tau) + \partial_\xi f_j(w_1(\xi,\tau), w_2(\xi,\tau),\ldots,w_N(\xi,\tau)) = 0,\\
&w_j(\xi,0) = u_{j,K}^{(n-1)}, \text{if } \xi<0,\\
&w_j(\xi,0) = u_{j,L}^{(n-1)}, \text{if } \xi>0, \qquad j=1,\ldots,N.
\end{aligned}$$

The Godunov scheme is a very efficient scheme when the above solution of the Riemann problem is easily constructed. When the solution of the Riemann problem becomes too complex or expensive, alternative schemes based on approximate Riemann solvers can be used.

References

- Roubíček, T. (2013), Nonlinear Partial Differential Equations with Applications (2nd ed.), Basel, Boston, Berlin: Birkhäuser, doi:10.1007/978-3-0348-0513-1, ISBN 978-3-0348-0512-4, MR 3014456
- Sarra, Scott (2003), "The Method of Characteristics with applications to Conservation Laws", Journal of Online Mathematics and its Applications
- Finite-volume-method: scholarpedia.org, Retrieved 25 June, 2020
- Subramanian, V.R.; R.E. White (2004), "Semianalytical method of lines for solving elliptic

partial differential equations", Chemical Engineering Science, 59 (4): 781–788, doi:10.1016/j.ces.2003.10.019

- Lee, H. J.; W. E. Schiesser (2004). Ordinary and Partial Differential Equation Routines in C, C++, Fortran, Java, Maple and Matlab. CRC Press. ISBN 1-58488-423-1

5

Applications in Various Fields

Differential equations are used in a wide variety of disciplines, from biology, economics, physics, chemistry and engineering. They are used in solving growth and decay problems, temperature problems, dilution problems, electrical circuits, etc. This chapter delves into the varied uses of differential equations for a thorough understanding of the subject.

ORTHOGONAL TRAJECTORIES

The term orthogonal means perpendicular, and trajectory means path or curve. Orthogonal trajectories, therefore, are two families of curves that always intersect perpendicularly. A pair of intersecting curves will be perpendicular if the product of their slopes is −1, that is, if the slope of one is the negative reciprocal of the slope of the other. Since the slope of a curve is given by the derivative, two families of curves $f_1(x, y, c) = 0$ and $f_2(x, y, c) = 0$ (where c is a parameter) will be orthogonal wherever they intersect if:

$$\frac{df_2}{dx} = \frac{-1}{df_1 / dx}$$

Example: The electrostatic field created by a positive point charge is pictured as a collection of straight lines which radiate away from the charge as shown in figure. Using the fact that the equipotentials (surfaces of constant electric potential) are orthogonal the electric field lines, determine the geometry of the equipotenitials of a point charge.

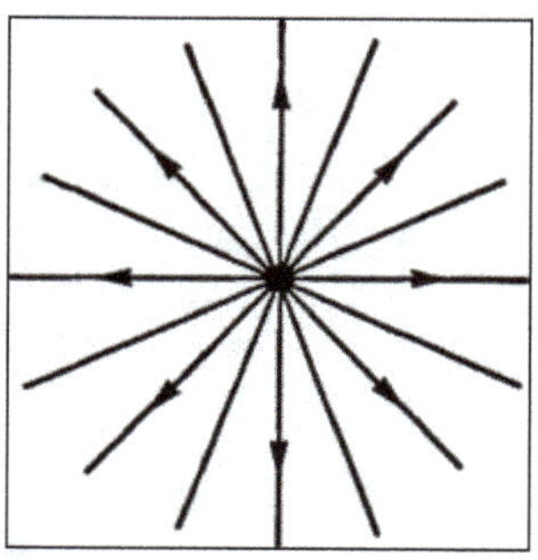

If the origin of an xy coordinate system is placed at the charge, then the electric field lines can be described by the family:

$$y = cx$$

The first step in determining the orthogonal trajectories is to obtain an expression for the slope of the curves in this family that does not involve the parameter c. In the present case:

$$y = cx \Rightarrow \frac{dy}{dx} = c = \frac{y}{x} \quad (*)$$

The differential equation describing the orthogonal trajectories is therefore:

$$\frac{dy}{dx} = -\frac{x}{y} \quad (**)$$

Since the right-hand side of (**) is the negative reciprocal of the right-hand side of (*). Because this equation is separable, the solution can proceed as follows:

$$\frac{dy}{dx} = -\frac{x}{y}$$

$$y\,dy = -x\,dx$$

$$\int y\,dy = \int -x\,dy$$

$$\frac{1}{2}y^2 = -\frac{1}{2}x^2 + c'$$

$$x^2 + y^2 = c^2$$

Where, c 2 = 2 c′.

The equipotential lines (that is, the intersection of the equipotential surfaces with any plane containing the charge) are therefore the family of circles $x^2 + y^2 = c^2$ centered at the origin. The equipotential and electric field lines for a point charge are shown in figure.

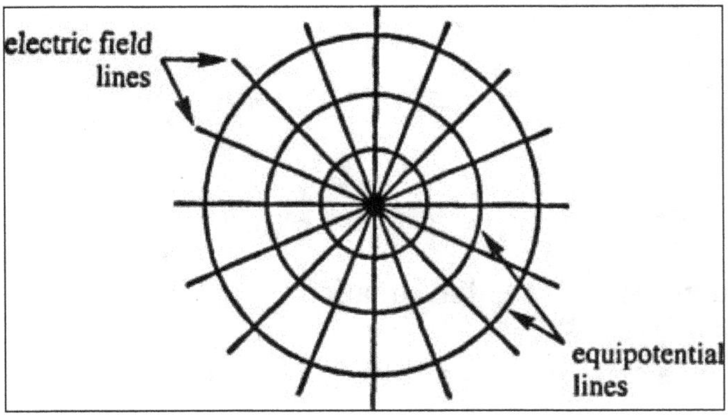

Example: Determine the orthogonal trajectories of the family of circles $x^2 + (y - c)^2 = c^2$ tangent to the x axis at the origin.

The first step is to determine an expression for the slope of the curves in this family that does not involve the parameter c. By implicit differentiation:

$$x^2 + (y-c)^2 = c^2$$
$$x^2 + y^2 = 2cy$$
$$2x + 2y\frac{dy}{dx} = 2c\frac{dy}{dx}$$
$$\frac{dy}{dx} = \frac{x}{c-y}$$

To eliminate c, note that:

$$x^2 + (y-c)^2 = c^2 \Rightarrow x^2 + y^2 = 2cy \Rightarrow c = \frac{x^2 + y^2}{2y}$$

The expression for dy/dx may now be written in the form,

$$\frac{dy}{dx} = \frac{x}{c-y} = \frac{x}{\frac{x^2+y^2}{2y} - y} = \frac{2xy}{x^2 - y^2} \quad (*)$$

Therefore, the differential equation describing the orthogonal trajectories is:

$$\frac{dy}{dx} = \frac{x^2 - y^2}{2xy} \quad (**)$$

Since the right-hand side of (**) is the negative reciprocal of the right-hand side of (*).

If equation (**) is written in the form:

$$(x^2 - y^2)dx - 2xy\,dy = 0$$

Note that it is not exact (since $M_y = 2y$ but $N_x = -2y$). However, because:

$$\frac{M_y - N_x}{N} = \frac{2y - (-2y)}{-2xy} = -\frac{2}{x}$$

is a function of x alone, the differential equation has:

$$\mu = e^{\int(-2/x)dx} = e^{-2\ln x} = e^{\ln(x^{-2})} = x^{-2}$$

as an integrating factor. After multiplying through by $\mu = x^{-2}$, the differential equation describing the desired family of orthogonal trajectories becomes:

$$\left(x^{-2}y^2 - 1\right)dx - 2x^{-1}ydy = 0$$

which is now exact (because $M_y = 2x^{-2}y = N_x$). Since:

$$\int \overline{M}\partial x = \int\left(x^{-2}y^2 - 1\right)\partial x = -x^{-1}y^2 - x$$

and,

$$\int \overline{N}\partial y = \int\left(-2x^{-1}y\right)\partial y = -x^{-1}y^2$$

the solution of the differential equation is:

$$-x^{-1}y^2 - x = -2c$$

(The reason the constant was written as $-2c$ rather than as c will be apparent in the following calculation.) With a little algebra, the equation for this family may be rewritten:

$$-x^{-1}y^2 - x = -2c$$

$$x^{-1}y^2 + x = 2c$$

$$y^2 + x^2 = 2cx$$

$$\left(x - c\right)^2 + y^2 = c^2$$

This shows that the orthogonal trajectories of the circles tangent to the x axis at the origin are the circles tangent to the y axis at the origin.

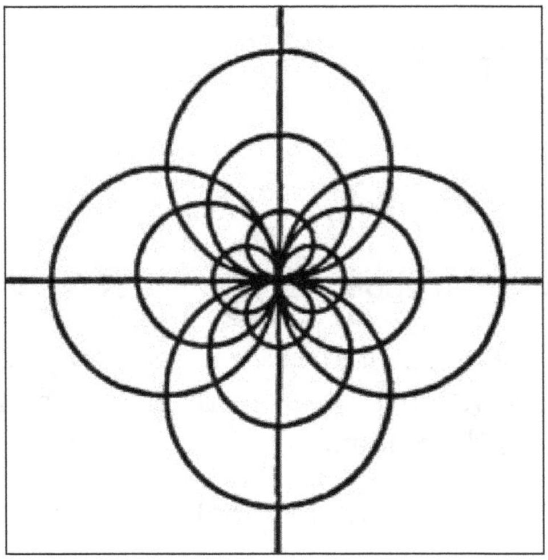

GROWTH AND DECAY PROBLEMS

Let N(t) denote amount of substance (or population) that is either growing or decaying. If we assume that $\frac{dN}{dt}$, the time rate of change of this amount of substance, is proportional to the amount of substance present, then $\frac{dN}{dt}=kN$ where k is the constant of proportionality.

When k is positive, population grows, when k is negative - population decays.

We are assuming that N(t) is a differentiable, hence continuous, function of time. For population problems, where N(t) is actually discrete and integer-valued, this assumption is incorrect. Nonetheless, above model still provides a good approximation to the physical laws governing such a system.

- This differential equation is separable. Rewriting it we obtain $\frac{dN}{N}=kdt$.
- Integration of both sides gives: $\ln(N)=kt+c_1$ or $N=Ce^{kt}$ where $C=e^{c_1}$.
- If we are given that $N(0)=N_0$ then $N_0=Ce^{k.0}$ or $C=n_0$.
- So, $N=N_0e^{kt}$.

The population of a certain country is known to increase at a rate proportional to the number of people presently living in the country. If after two years the population has doubled, and after three years the population is 20000, estimate the number of people initially living in the country.

- Let N_0 is number of people initially living in the country, then $N(0)=N_0, N(2)=2N_0, N(3)=20000$.
- So, as was if $N(0)=N_0$ then $N=N_0e^{kt}$.
- Next, since $N(2)=2N_0$ then $2N_0=N_0e^{k.2}$ or $2k=\ln(2)$. This gives that $k=\frac{\ln(2)}{2}$.
- Now, equation can be rewritten as $N=N_0e^{\frac{\ln(2)}{2}t}$.
- Since $N(3)=20000$ then $20000=N_0e^{\frac{\ln(2)}{2}3}$ or $N_0=20000e^{-1.5\ln(2)}\approx 7071$ people.

A certain radioactive material is known to decay at a rate proportional to the amount present. If after one hour it is observed that 10 percent of the material has decayed, find the half-life (period of time it takes for the amount of material to decrease by half) of the material.

- Let initial amount of radioactive material is N_0 then $N(0)=N_0, N(1)=(1=0.1)N =$ $N_0=0.9N_0$.

- We need to find such t that $N(t)=\frac{N_0}{2}$.
- So, as was, if $N(0)=N_0$ then $N(t)=N_0e^{kt}$.
- Since $N(1)= 0.9N_0$ then $0.9N_0=N_0e^{k.1}$ or $k=\ln(0.9)$.
- Now equation has form $N(t)=N_0e^{\ln(0.9)t}$.
- Now, find t, such that $N(t)=0.5N_0 : 0.5N_0=N_0e^{\ln(0.9)t}$ or $t=\frac{\ln(0.5)}{\ln(0.9)}\approx 6.58$ hours.

TEMPERATURE PROBLEMS

Newton's law of cooling, which is equally applicable to heating, states that the time rate of change of the temperature of a body is proportional to the temperature difference between the body and its surrounding medium. Let T denote the temperature of the body and let T_m denote the temperature of the surrounding medium. Then the time rate of change of the temperature of the body is $\frac{dT}{dt}$, and Newton's law of cooling can be formulated as $\frac{dT}{dt}=-k(T-T_m)$ or as $\frac{dT}{dt}+kT=kT_m$ where k is a positive constant of proportionality. Once k is chosen positive, the minus sign is required in Newton's law to make $\frac{dT}{dt}$ negative in a cooling process, when T is greater than T_m, and positive in a heating process, when T is less than T_m.

A body at an unknown temperature is placed in a room which is held at a constant temperature of 30° F. If after 10 minutes the temperature of the body is 0° F and after 20 minutes the temperature of the body is 15° F, find the unknown initial temperature.

- We have that $T_m = 30$, so differential equation is $\frac{dT}{dt}+kT=30k$. This is first-order linear differential equation. Integrating factor is $I=e^{\int kdt}=e^{kt}$. After multiplying equation by integrating factor we obtain that $e^{kt}\frac{dT}{dt}+kTe^{kt}=30ke^{kt}$ or $\frac{d(Te^{kt})}{dt}=30ke^{kt}$.
- Integrating both sides gives $Te^{kt} = 30e^{kt}+C$ or $T=Ce^{-kt}+30$.
- We are given that $T(10)=0$ or $0=Ce^{-10k}+30$. Also $T(20)=15$ or $15=Ce^{-20k}+30$.

Thus, we have system of two equations:

$$\begin{cases} Ce^{-10k} = -30 \\ Ce^{-20k} = -15 \end{cases}$$

Dividing first equation by second gives $e^{10k} = 2$. Now, from first equation $C = -30e^{10k} = -30 \cdot 2 = -60$.

Finally, $T_0 = T(0) = Ce^{-k \cdot 0} + 30 = C + 30 = -60 + 30 = -30$.

A body at a temperature of 50° F is placed in an oven whose temperature is kept at 150° F. If after 10 minutes the temperature of the body is 75° F, find the time required for the body to reach a temperature of 100° F.

- We have that $T(0) = 50, T(10) = 75, T_m = 150$.
- So, differential equation is $\frac{dT}{dt} + kT = 150k$. This is again as in example 1 is linear first-order differential equation. Its solution is $T = Ce^{-kt} + 150$.
- Since $T(0) = 50$ then $50 = Ce^{-k \cdot 0} + 150$ or $C = -100$.

Now equation has form $T = -100e^{-kt} + 150$;

- Since $T(10) = 75$ then $75 = -100e^{-10k} + 150$ or $k = -\frac{\ln(0.75)}{10}$.
- Finally equation has the following form: $T = -100e^{\frac{\ln(0.75)}{10}} + 150$.
- Now, we need to find such t that $T(t) = 100$:
- $100 = -100e^{\frac{\ln(0.75)}{10}} + 150$ or $t = 10\frac{\ln(0.5)}{\ln(0.75)} \approx 24.0942$ minutes.

FALLING BODY PROBLEMS

Consider a vertically falling body of mass m that is being influenced only by gravity g and an air resistance that is proportional to the velocity of the body. Assume that both gravity and mass remain constant and, for convenience, choose the downward direction as the positive direction.

Newton's second law of motion: The net force acting on a body is equal to the time rate of change of the momentum of the body; or, for constant mass, $F = m\frac{dv}{dt}$.

For the problem at hand, there are two forces acting on the body: the force due to gravity given by the weight w of the body, which equals mg, and the force due to air resistance given by $-kv$, where $k>0$ is a constant of proportionality. The minus sign is required because this force opposes the velocity; that is, it acts in the upward, or negative, direction. The net force F on the body is, therefore, $F = mg-kv$.

Substituting this result we obtain $mg-kv=m\frac{dv}{dt}$ or $\frac{dv}{dt}+\frac{k}{m}v=g$ as the equation of motion for the body.

If air resistance is negligible or nonexistent, then k = 0 and equation simplifies to $\frac{dv}{dt}=g$.

When k > 0, the limiting velocity v_1 is defined by $v_1=\frac{mg}{k}$.

Given differential equation is valid only if the given conditions are satisfied. This equation is not valid if, for example, air resistance is not proportional to velocity but to the velocity squared, or if the upward direction is taken to be the positive direction.

A steel ball weighing 2 Ib is dropped from a height of 3000 ft with no velocity. As it falls, the ball encounters air resistance numerically equal to $\frac{v}{8}$ (in pounds), where v denotes the velocity of the ball (in feet per second). Find the limiting velocity for the ball and the time required for the ball to hit the ground.

Locate the coordinate system with the ground now situated at $x=3000$. Here, w=2 lb and $k=\frac{1}{8}$. Assuming gravity g is $32\frac{ft}{s^2}$, we have from the formula $w=mg$ that $m=\frac{w}{g}=\frac{2}{32}=\frac{1}{16}$ slug. So, differential equation is $\frac{dv}{dt}+\frac{\frac{1}{8}}{\frac{1}{16}}v=32$ or $\frac{dv}{dt}+2v=32$.

This is first-order linear differential equation. Its solution is $v(t)=ce^{-2t}+16$.

We are given that initially there is no velocity, so $v(0)=0:0=ce^{-2\cdot 0}+16$ or $c=-16$.

And differential equation becomes $v(t)=-16e^{-2t}+16$.

From this equation we see that $v\to 16$ when $t\to\infty$, so limiting velocity is $16\frac{ft}{s^2}$.

To find the time it takes for the ball to hit the ground (x = 3000), we need an expression for the position of the ball at any time t. Since $v=\frac{dx}{dt}$ then $\frac{dx}{dt}=-16e^{-2t}+16$. Integrating both sides gives $x=8e^{-2t}+16t+c_1$. To find constant use the fact that $x=0$ when $t=0:0=8e^{-2\cdot 0}+16\cdot 0+c_1$ or $c_1=-8$.

So, position of particle is given as $x(t)=8e^{-2t}+16t-8$.

We need to find such t that $x(t)=3000: 3000=8e^{-2t}+16t-8$ or $376=e^{-2t}+2t$.

This equation cannot be solved explicitly for t, we can approximate the solution by trial and error, substituting different values of t until we locate a solution to the degree of accuracy we need. Alternatively, we note that for any large value of t, the negative exponential term will be negligible. A good approximation is obtained by setting 2t = 376 or t = 188 sec. For this value of t, the exponential is essentially zero.

A body is propelled straight up with an initial velocity of 500 ft/sec in a vacuum with no air resistance. How long will it take the body to return to the ground?

Since there is no air resistance then k=0, so $\frac{dv}{dt}=g=32$, so $v=32t+C$.

Since $v(0)=-500$ (because we are moving up in negative direction) then $-500=32\cdot 0+C$ or $C=-500$.

So, $v=32t-500$.

It is known that $\frac{dx}{dt}=v=32t-500$. Integrating this equation gives $x(t)=16t^2-500t+C_1$.

Since $x(0)=0$ then $0=16\cdot 0^2-500\cdot 0+C_1$ or $C_1=0$.

So, $x(t)=16t^2-500t$. We need to find time when $x(t)=0$.

$16t^2-500t=0$ or $t(16t-500)=0$ which gives $t=\frac{500}{16}=31.25$ seconds.

DILUTION PROBLEMS

Consider a tank which initially holds V_0 gal of brine that contains a lb of salt. Another brine solution, containing b lb of salt per gallon, is poured into the tank at the rate of e gal/min while, simultaneously, the well-stirred solution leaves the tank at the rate of f gal/min. The problem is to find the amount of salt in the tank at any time t.

Let Q denote the amount (in pounds) of salt in the tank at any time t. The time rate of change of Q, $\frac{dQ}{dt}$, equals the rate at which salt enters the tank minus the rate at which salt leaves the tank. Salt enters the tank at the rate of be lb/min. To determine the rate at which salt leaves the tank, we first calculate the volume of brine in the tank at any time t, which is the initial volume V_0 plus the volume of brine added et minus the volume of brine removed ft . Thus, the volume of brine at any time is $V_0+et-ft$. The concentration of salt in the tank at any time is $\frac{Q}{V_0+et-ft}$, from which it follows that salt leaves the tank at the rate of $f\frac{Q}{V_0+et-ft}$ gal/min.

Thus, $\frac{dQ}{dt}=be-f\frac{Q}{V_o+et-ft}$ or $\frac{dQ}{dt}+f\frac{f}{V_o+(e-f)t}Q=be.$

A tank initially holds 100 gal of a brine solution containing 20 Ib of salt. At t = 0, fresh water is poured into the tank at the rate of 5 gal/min, while the well-stirred mixture leaves the tank at the same rate. Find the amount of salt in the tank at any time t.

Here, $V_o=100, a=20, b=0$ and $e=f=5$. So, equation is $\frac{dQ}{dt}+\frac{1}{20}Q=0$. Solution to this linear equation is $Q(t)=Ce^{-\frac{t}{20}}$.

So, $Q(t)=20e^{-\frac{t}{20}}$.

Note that $Q\to 0$ when $t\to\infty$ as it should, since only fresh water is being added.

A tank contains 40 l of solution containing 2 g of substance per liter. Salt water containing 3 g of this substance per liter runs in at the rate of 4 l/min and the well-stirred mixture runs out at the same rate. Find the amount of substance in the tank after 15 minutes.

Note that here we have problem in terms of liters and grams, not in terms of pounds and gallons. It doesn't matter. The main thing is that all variables were presented using same units.

Since initially there are 40 l of solution and each liter contains 2 grams of substance then initially there are 40*2=80 grams of substance.

So, $V_o=40 a=80, b=3, e=f=4$.

And equation is written as $\frac{dQ}{dt}+\frac{1}{10}Q=12$. This is linear equation. Its solution is $Q(t)=120+Ce^{-\frac{t}{10}}$.

At t=0, we are given that $Q=a=80$, so $80=120+Ce^{-\frac{0}{10}}$ or $C=-40$.

Thus, $Q(t)=120-40e^{-\frac{t}{10}}$.

We need to find $Q(15): Q(15)=120-40e^{-\frac{15}{10}}\approx 111.075$ grams.

ELECTRICAL CIRCUITS

The basic equation governing the amount of current I (in amperes) in a simple RL circuit consisting of a resistance R (in ohms), an inductor L (in henries), and an electromotive force (abbreviated emf) E (in volts) is $\frac{dI}{dt}+\frac{R}{L}I=\frac{E}{L}$.

For an RC circuit consisting of a resistance, a capacitance C (in farads), an emf, and no inductance, the equation governing the amount of electrical charge q (in coulombs) on the capacitor is $\frac{dq}{dt}+\frac{1}{RC}q=\frac{E}{R}$. The relationship between q and I is $I=\frac{dq}{dt}$.

An RL circuit has an emf of 5 volts, a resistance of 50 ohms, an inductance of 1 henry, and no initial current. Find the current in the circuit at any time t.

Here E=5, R=50, and L=1; hence differential equation becomes $\frac{dI}{dt}+50I=5$. This is linear equation, its solution is $I=ce^{-50t}+\frac{1}{10}$.

At t=0, I=0; th us, $0=ce^{-50\cdot0}+\frac{1}{10}$ or $c=-\frac{1}{10}$. The current at any time, then, $I=-\frac{1}{10}e^{-50t}+\frac{1}{10}$.

The quantity $-\frac{1}{10}e^{-50t}$ is called the transient current, since this quantity goes to zero ("dies out") as $t\to\infty$.

The quantity $\frac{1}{10}$ is called the steady-state current. As $t\to\infty$, the current I approaches the value of the steady-state current.

A RC circuit has an emf of 300cos(2t) volts, a resistance of 150 ohms, a capacitance of $\frac{1}{600}$ farad, and an initial charge on the capacitor of 5 coulombs. Find the charge on the capacitor at any time t and the steady-state current.

Here $E=300\cos(2t), R=150, C=\frac{1}{600}$.

So, $\frac{dq}{dt}+4q=2\cos(2t)$. This is linear differential equation. It can be rewritten as $\frac{dq}{dt}e^{4t}+4e^{4t}q=2e^{4t}\cos(2t)$ or $\frac{d(qe^{4t})}{dt}=2e^{4t}\cos(2t)$.

Integrating both sides gives $qe^{4t}=\int(2e^{4t}\cos(2t))dt$. Using integration by parts we obtain that $qe^{4t}=c+\frac{e^{4t}}{5}(\sin(2t)+2\cos(2t))$.

Charge on the capacitor at any time t is $q=ce^{-4t}+\frac{1}{5}(\sin(2t)+2\cos(2t))$.

Since q=5 when t=0 then $5=ce^{-4\cdot0}+\frac{1}{5}(\sin(2\cdot0)+2\cos(2\cdot0))$ or $c=\frac{23}{5}$.

Thus, $q=\frac{1}{5}(23e^{-4t}+\sin(2t)+2\cos(2t))$.

Next since, $I=\frac{dq}{dt}$ then $I=\frac{1}{5}\left(-92e^{-4t}+2\cos(2t)-4\sin(2t)\right)$ and the steady-state current is $I_s=\frac{1}{5}(2\cos(2t)-4\sin(2t))$ $\left(-\frac{92}{5}e^{-4t}\right.$ is transient because this quantity goes to zero as $t\to\infty$, unlike $\left.I_s\right)$.

VELOCITY AND ACCELERATION IN SIMPLE HARMONIC MOTION

A motion is said to be accelerated when its velocity keeps changing. But in simple harmonic motion, the particle performs the same motion again and again over a period of time.

Acceleration in SHM

We know what acceleration is. It is velocity per unit time. We can calculate the acceleration of a particle performing S.H.M. The differential equation of linear S.H.M. is $d^2x/dt^2+(k/m)x=0$ where d^2x/dt^2 is the acceleration of the particle, x is the displacement of the particle, m is the mass of the particle and k is the force constant. We know that $k/m=\omega^2$ where ω is the angular frequency.

Therefore, $d^2x/dt^2+\omega^2x=0$.

Hence, acceleration of S.H.M.= $d^2x/dt^2=-\omega^2x$.

The negative sign indicated that acceleration and displacement are in opposite direction of each other. Equation I is the expression of acceleration of S.H.M. Practically, the motion of a particle performing S.H.M. is accelerated because its velocity keeps changing either by a constant number or varied number.

Take a simple pendulum for example. When we swing a pendulum, it moves to and fro about its mean position. But after some time, it eventually stops and returns to its mean position. This type of simple harmonic motion in which velocity or amplitude keeps changing is damped simple harmonic motion.

Velocity in SHM

Velocity is distance per unit time. We can obtain the expression for velocity using the expression for acceleration. Acceleration $d^2x/dt^2=dv/dt=dv/dx\times dx/dt$. But $dx/dt=\text{velocity}\ 'v'$.

Therefore, acceleration $= v(dv/dx)$.

When we substitute equation $v(dv/dx)$ in $d^2x/dt^2 = -\omega^2 x$.

$$\therefore vdv = -\omega^2 xdx$$

After integrating both sides, we get,

$$\int vdv = \int -\omega^2 xdx = -\omega^2 \int xdx$$

Hence, $v^2/2 = -\omega^2 x^2/2 + C$ where C is the constant of integration. Now, to find the vaue of C, consider boundary value condition. When a particle performing SHM is at the extreme position, displacement of the particle is maximum and velocity is zero. (a is the amplitude of SHM),

Therefore, at $x = \pm a$, $v = 0$.

And $0 = -\omega^2 a^2/2 + C$.

Hence, $C = \omega^2 a^2/2$.

Substitute this value of C in equation $v^2/2 = -\omega^2 x^2/2 + C$.

$$\therefore v^2/2 = -\omega^2 x^2/2 + \omega^2 a^2/2.$$

$$\therefore v^2 = \omega^2 \left(a^2 - x^2\right)$$

Taking square root on both sides, we get,

$$v = \pm \omega \sqrt{\left(a^2 - x^2\right)}$$

The above equation is the expression of the velocity of S.H.M. The double sign indicates that when a particle passes through a given point in the positive direction of x, v is positive, and when it passes through the same point in opposite direction of x, v is negative.

Maximum and Minimum Velocity

We know the velocity of a particle performing S.H.M. is given by, $v = \pm \omega \sqrt{a^2 - x^2}$. At mean position, x = 0. Therefore, $v = \pm \omega \sqrt{a^2 - 0^2} = \pm \omega \sqrt{a^2} = \pm a\omega$. Therefore, at mean position, velocity of the particle performing S.H.M. is maximum which is $V_{max} = \pm a\omega$. At extreme position, $x = \pm a$.

Therefore, $v = \pm \omega \sqrt{a^2 - a^2} = \omega \times 0 = 0$. Therefore, at extreme position, velocity of the particle performing S.H.M. is minimum which is $V_{min} = 0$.

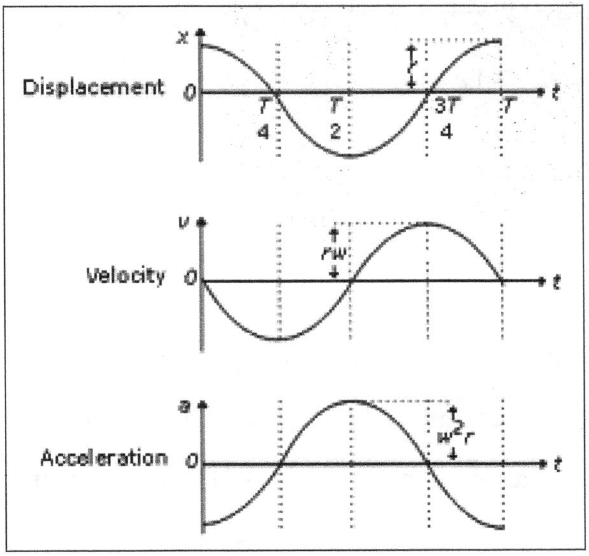

Example: What is the value of acceleration at the mean position?

Solution: At mean position, x =0.

$\therefore$ Acceleration $= -\omega^2 x = -\omega^2 \times 0$. Therefore, the value of acceleration at the mean position is minimum and it is zero.

MECHANICAL VIBRATIONS

A mass m is suspended at the end of a spring, its weight stretches the spring by a length L to reach a static state (the equilibrium position of the system). Let u(t) denote the displacement, as a function of time, of the mass relative to its equilibrium position. u > 0 means the spring is stretched beyond its equilibrium length, while u < 0 means that the spring is compressed. The mass is then assumed to be set in motion (by any one of several means).

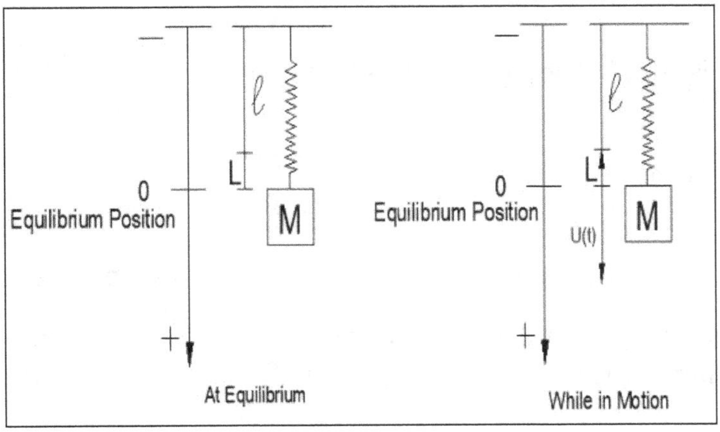

The Equations that Govern a Mass-spring System

At equilibrium: (by Hooke's Law)

$$mg = kL$$

While in motion:

$$mu'' + \gamma u' + ku = F(t)$$

This is a second order linear differential equation with constant coefficients. It usually comes with two initial conditions: $u(t_0) = u_0$, and $u'(t_0) = u'_0$.

- u(t) = displacement of the mass relative to its equilibrium position.
- m = mass (m > o).
- γ = damping constant (γ ≥ o).
- k = spring (Hooke's) constant (k > o).
- g = gravitational constant.
- L = elongation of the spring caused by the weight.
- F(t) = externally applied forcing function, if any.
- u(to) = initial displacement of the mass.
- u′(to) = initial velocity of the mass.
- Undamped Free Vibration $(\gamma = 0, F(t) = 0)$

The simplest mechanical vibration equation occurs when $\gamma = 0, F(t) = 0$. This is the undamped free vibration. The motion equation is:

$$mu'' + ku' = 0.$$

The characteristic equation is $mr^2 + k = 0$. Its solutions are $r = \pm\sqrt{\frac{k}{m}}i$. The general solution is then:

$$u(t) = C_1 \cos\omega_0 t + C_2 \sin\omega_0 t.$$

Where $\omega_0 = \sqrt{\frac{k}{m}}$ is called the natural frequency of the system. It is the frequency at which the system tends to oscillate in the absence of any damping. A motion of this type is called simple harmonic motion. It is a perpetual, sinusoidal, motion.

Just like everywhere else in calculus, the angle is measured in radians, and the (angular) frequency is given in radians per second. The frequency is not given in hertz (which measures the number of cycles or revolutions per second). Instead, their relation is: 2π radians/sec = 1 hertz.

The (natural) period of the oscillation is given by $T=\frac{2\pi}{\omega_0}$ (seconds).

To get a clearer picture of how this solution behaves, we can simplify it with trig identities and rewrite it as:

$$u(t)=R\cos(\omega_0 t-\delta).$$

The displacement is oscillating steadily with constant amplitude of oscillation:

$$R=\sqrt{C_1^{\,2}+C_2^{\,2}}.$$

The angle δ is the phase or phase angle of displacement. It measures how much u(t) lags (when $\delta > 0$), or leads (when $\delta < 0$) relative to $\cos(\omega_0 t)$, which has a peak at $t = 0$. The phase angle satisfies the relation:

$$\tan\delta=\frac{C_1}{C_2}.$$

More explicitly, it is calculated by:

$$\delta=\tan^{-1}\frac{C_2}{C_1},\ \text{if}\ C_1>0,$$

$$\delta=\tan^{-1}\frac{C_2}{C_1}+\pi,\ \text{if}\ C_1>0,$$

$$\delta=\frac{\pi}{2},\ \text{if}\ C_1=0\ \text{and}\ C_2>0,$$

$$\delta=-\frac{\pi}{2},\ \text{if}\ C_1=0\ \text{and}\ C_2<0,$$

The angle is undefined if $C_1 = C_2 = 0$.

An example of simple harmonic motion:

$\text{Graph of } u(t)=\cos(t)-\sin(t)$

$\text{Amplitude}: R=\sqrt{2}$

$\text{Phase angle}: \delta=-\pi/4$

Damped Free Vibration $(\gamma>0, F(t)=0)$

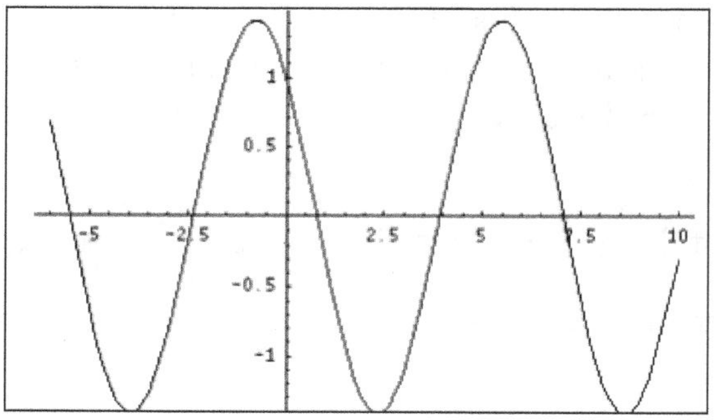

When damping is present (as it realistically always is) the motion equation of the unforced mass-spring system becomes:

$$mu''+\gamma u'+ku=0$$

Where m, γ, k are all positive constants. The characteristic equation is $mr^2+\gamma r+k=0$.

Its solution(s) will be either negative real numbers, or complex numbers with negative real parts. The displacement u(t) behaves differently depending on the size of γ relative to m and k. There are three possible classes of behaviors based on the possible types of root(s) of the characteristic polynomial.

Case I: Two distinct (negative) real roots.

When $\gamma^2>4mk$, there are two distinct real roots, both are negative. The displacement is in the form $u(t)=C_1e^{r_1t}+C_2e^{r_2t}$.

A mass-spring system with such type displacement function is called over damped. Note that the system does not oscillate; it has no periodic components in the solution. In fact, depending on the initial conditions the mass of an over damped mass-spring system might or might not cross over its equilibrium position. But it could cross the equilibrium position at most once.

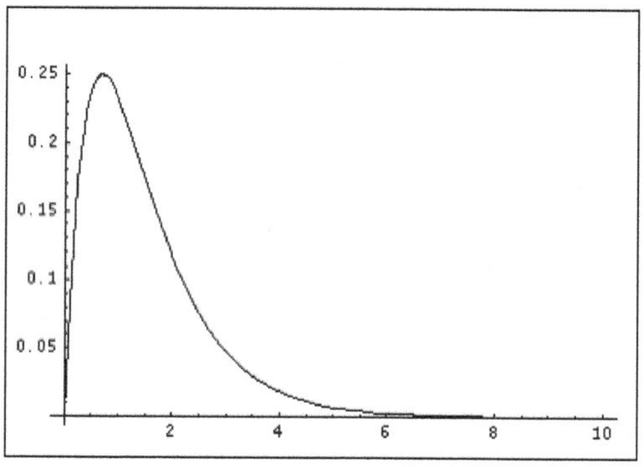

Graph of $u(t)=e^{-t}-e^{-2t}$.

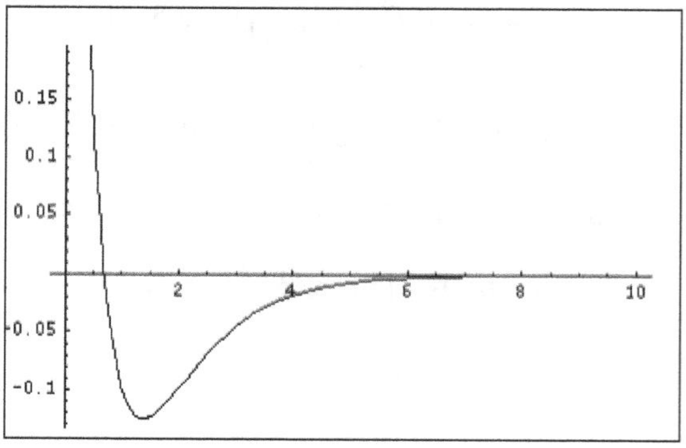

Graph of $u(t)=-e^{-t}+2e^{-2t}$.

Displacement of an over damped system.

Case II: One repeated (negative) real root.

When $\gamma^2=4mk$, there is one (repeated) real root. It is negative: $r=\frac{-\gamma}{2m}$.

The displacement is in the form:

$$u(t)=C_1e^{rt}+C_2te^{rt}.$$

A system exhibits this behavior is called critically damped. That is, the damping coefficient γ is just large enough to prevent oscillation. As can be seen, this system does not oscillate, either. Just like the over damped case, the mass could cross its equilibrium position at most one time.

The value $\gamma^2=4mk \rightarrow \gamma=2\sqrt{mk}$ is called critical damping. It is the threshold level below which damping would be too small to prevent the system from oscillating.

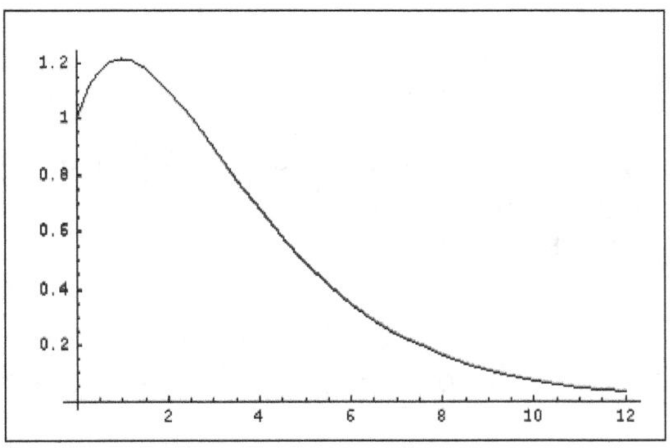

Graph of $u(t)=e^{-t/2}+te^{-t/2}$.

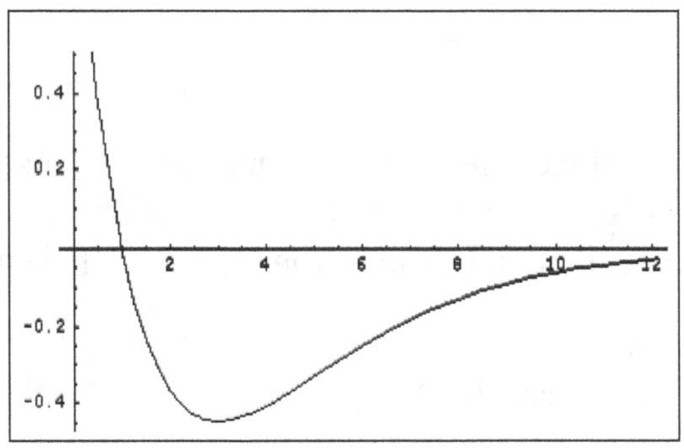

Graph of $u(t)=e^{-t/2}-te^{-t/2}$.

Displacement of a critically damped system.

Case III: Two complex conjugate roots.

When $\gamma^2<4mk$, there are two complex conjugate roots, where their common real part, λ is always negative. The displacement is in the form:

$$u(t)=C_1e^{\lambda t}\cos\mu t+C_2e^{\lambda t}\sin\mu t.$$

A system exhibits this behavior is called under damped. The name means that the damping is small compares to m and k, and as a result vibrations will occur. The system oscillates (note the sinusoidal components in the solution). The displacement function can be rewritten as:

$$u(t)=Re^{\lambda t}\cos(\mu t-\delta).$$

The formulas for R and δ are the same as in the previous (undamped free vibration) section. The displacement function is oscillating, but the amplitude of oscillation, $Re^{\lambda t}$ is decaying exponentially. For all particular solutions (except the zero solution that corresponds to the initial conditions $u(t_o)=0, u'(t_o)=0$), the mass crosses its equilibrium position infinitely often.

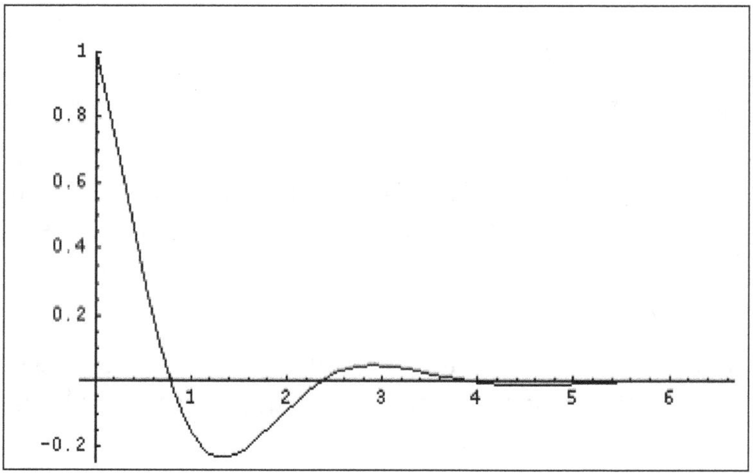

Damped oscillation: $u(t)=e^{-t}\cos(2t)$.

The displacement of an under damped mass-spring system is a quasi-periodic function (that is, it shows periodic-like motion, but it is not truly periodic because its amplitude is ever decreasing so it does not exactly repeat itself).

It is oscillating at quasi-frequency, which is μ radian per second. (It's just the frequency of the sinusoidal components of the displacement.) The peakto-peak time of the oscillation is the quasi-period: $T_q = \dfrac{2\pi}{\mu}$ (seconds).

In addition to cause the amplitude to gradually decay to zero, damping has another, more subtle, effect on the oscillating motion: It immediately decreases the quasi-frequency and, therefore, lengthens the quasi-period (compare to the natural frequency and natural period of an undamped system). The larger the damping constant γ, the smaller quasi-frequency and the longer the quasi-period become. Eventually, at the critical damping threshold, when $\gamma=\sqrt{4mk}$, the quasi-frequency vanishes and the displacement becomes aperiodic (becoming instead a critically damped system).

Note that in all 3 cases of damped free vibration, the displacement function tends to zero as $t \to \infty$. This behavior makes perfect sense from a conservation of energy point-of-view: while the system is in motion, the damping wastes away whatever energy the system has started out with, but there is no forcing function to supply the system with additional energy. Consequently, eventually the motion comes to a halt.

Example: A mass of 1 kg stretches a spring 0.1 m. The system has a damping constant of $\gamma = 14$. At t = 0, the mass is pulled down 2 m and released with an upward velocity of 3.5 m/s. Find the displacement function. What are the systems' quasi-frequency and quasi-period?

$$m=1, \gamma=14, L=0.1;$$

$$mg=9.8=kL=0.1k \rightarrow 98=k.$$

The motion equation is $u''+14u'+98u=0$, and the initial conditions are $u(0)=2, u'(0)=-3.5$.

The roots of characteristic polynomial are $r=-7\pm 7i$:

$$u(t)=C_1e^{-7t}\cos 7t+C_2e^{-7t}\sin 7t$$

Therefore, the quasi-frequency is 7 (rad/sec) and the quasi-period is: $T_q=\frac{2\pi}{7}$ (seconds).

Apply the initial condition and we get $C_1=2$, and $C_2=3/2$. Hence, $u(t)=2e^{-7t}\cos 7t+1.5e^{-7t}\sin 7t$.

References

- Growth-and-decay-problems, applications-of-first-order-ode, differential-equations: emathhelp.net, Retrieved 26 May, 2020
- Applications-of-first-order-equations, applying-differential-equations, differential-equations, study-guides: cliffsnotes.com, Retrieved 19 July, 2020
- Temperature-problems, applications-of-first-order-ode, differential-equations: emathhelp.net, Retrieved 19 July, 2020
- Velocity-and-acceleration-in-simple-harmonic-motion, oscillations, physics, guides: toppr.com, Retrieved 16 August, 2020
- Differential-equations, 8-2nd-order-de-damping-rlc: intmath.com, Retrieved 10 June, 2020

Permissions

All chapters in this book are published with permission under the Creative Commons Attribution Share Alike License or equivalent. Every chapter published in this book has been scrutinized by our experts. Their significance has been extensively debated. The topics covered herein carry significant information for a comprehensive understanding. They may even be implemented as practical applications or may be referred to as a beginning point for further studies.

We would like to thank the editorial team for lending their expertise to make the book truly unique. They have played a crucial role in the development of this book. Without their invaluable contributions this book wouldn't have been possible. They have made vital efforts to compile up to date information on the varied aspects of this subject to make this book a valuable addition to the collection of many professionals and students.

This book was conceptualized with the vision of imparting up-to-date and integrated information in this field. To ensure the same, a matchless editorial board was set up. Every individual on the board went through rigorous rounds of assessment to prove their worth. After which they invested a large part of their time researching and compiling the most relevant data for our readers.

The editorial board has been involved in producing this book since its inception. They have spent rigorous hours researching and exploring the diverse topics which have resulted in the successful publishing of this book. They have passed on their knowledge of decades through this book. To expedite this challenging task, the publisher supported the team at every step. A small team of assistant editors was also appointed to further simplify the editing procedure and attain best results for the readers.

Apart from the editorial board, the designing team has also invested a significant amount of their time in understanding the subject and creating the most relevant covers. They scrutinized every image to scout for the most suitable representation of the subject and create an appropriate cover for the book.

The publishing team has been an ardent support to the editorial, designing and production team. Their endless efforts to recruit the best for this project, has resulted in the accomplishment of this book. They are a veteran in the field of academics and their pool of knowledge is as vast as their experience in printing. Their expertise and guidance has proved useful at every step. Their uncompromising quality standards have made this book an exceptional effort. Their encouragement from time to time has been an inspiration for everyone.

The publisher and the editorial board hope that this book will prove to be a valuable piece of knowledge for students, practitioners and scholars across the globe.

Index

www.ingramcontent.com/pod-product-compliance
Lightning Source LLC
LaVergne TN
LVHW080615210926
848485LV00003B/164
* 9 7 8 1 6 3 9 8 7 1 6 2 9 *